Object-Oriented Design
for Temporal GIS

Object-Oriented Design for Temporal GIS

MONICA WACHOWICZ

CRC Press
Taylor & Francis Group
Boca Raton London New York

CRC Press is an imprint of the
Taylor & Francis Group, an **informa** business
A TAYLOR & FRANCIS BOOK

First published 1999 by Taylor & Francis

Published 2019 by CRC Press
Taylor & Francis Group
6000 Broken Sound Parkway NW, Suite 300
Boca Raton, FL 33487-2742

First issued in paperback 2020

ISBN-13: 978-0-367-57917-3 (pbk)
ISBN-13: 978-0-7484-0831-3 (hbk)

Visit the Taylor & Francis Web site at
http://www.taylorandfrancis.com

and the CRC Press Web site at
http://www.crcpress.com

British Library Cataloging in Publication Data
A catalogue record for this book is available from the British Library.
ISBN 0-7484-0831-2 (cased)

Library of Congress Cataloging-in-Publication Data are available

Cover design by Hybert Design and Type, Waltham St Lawrence, Berkshire
Typeset in Times 10/12pt by Graphicraft Ltd, Hong Kong

Contents

Series Introduction

Welcome

The *Research Monographs in Geographical Information Systems* series provides a publication outlet for research of the highest quality in GIS, which is longer than would normally be acceptable for publication in a journal. The series includes single- and multiple-author research monographs, often based upon PhD theses and the like, and special collections of thematic papers.

The need

We believe that there is a need, from the point of view of both readers (researchers and practitioners) and authors, for longer treatments of subjects related to GIS than are widely available currently. We feel that the value of much research is actually devalued by being broken up into separate articles for publication in journals. At the same time, we realise that many career decisions are based on publication records, and that peer review plays an important part in that process. Therefore a named editorial board supports the series, and advice is sought from them on all submissions.

Successful submissions will focus on a single theme of interest to the GIS community, and treat it in depth, giving full proofs, methodological procedures or code where appropriate to help the reader appreciate the utility of the work in the monograph. No area of interest in GIS is excluded, although material should demonstrably advance thinking and understanding in spatial information science. Theoretical, technical and application-oriented approaches are all welcomed.

The medium

In the first instance the majority of monographs will be in the form of a traditional text book, but, in a changing world of publishing, we actively encourage publication on CD-ROM, the placing of supporting material on web sites, or publication of programs and of data. No form of dissemination is discounted, and prospective authors are invited to suggest whatever primary form of publication and support material they think is appropriate.

The editorial board

The monograph series is supported by an editorial board. Every monograph proposal is sent to all members of the board which includes Ralf Bill, António Câmara, Joseph Ferreira, Pip Forer, Andrew Frank, Gail Kucera, Enrico Puppo, and Peter van Oostrom. These people have been invited for their experience in the field, of monograph writing, and for their geographic and subject diversity. Members may also be involved later in the process with particular monographs.

Future submissions

Anyone who is interested in preparing a Research monograph, should contact either of the editors. Advice on how to proceed will be available from them, and is treated on a case by case basis.

For now we hope that you find this, the fifth in the series, a worthwhile addition to your GIS bookshelf, and that you may be inspired to submit a proposal too.

Editors:

Professor Peter Fisher
Department of Geography
University of Leicester
Leicester
LE1 7RH
UK
Phone: +44 (0) 116 252 3839
Fax: +44 (0) 116 252 3854
Email: pff1@le.ac.uk

Dr Jonathan Raper
Department of Information Science
City University
Northampton Square
London
EC1V 0HB
UK
Phone: +44 (0) 171 477 8415
Fax: +44 (0) 171 477 8584

Preface

A wide spectrum of research areas have contributed to the development of object-oriented analysis and design methods. Within computing sciences object-orientation has been developed in the fields of databases, programming languages, and system engineering. Some indications of this development include the proliferation of object-oriented concepts, notations, models, and application areas. This book provides a much-needed geographic perspective to enhance both the development and understanding of object-oriented methodology. In particular, it addresses how both Time Geography and the object-oriented methodologies can be integrated to represent and manipulate space and time in GIS.

Time Geography provides a conceptual framework for capturing the semantics of space and time. This book introduces the reader to the framework abstractions developed in Time Geography. They are of primordial importance for handling spatio-temporal data in GIS. They play an important role as a modelling tool for representing the passage of time and the mechanisms of change. The book explains how the time geographic abstractions can be integrated with the modelling constructs of object-oriented analysis and design methods. This is achieved by describing a spatio-temporal data model designed to address the complex and subtle semantics of space and time of historical data. The logical components (schema evolution, data update procedures, events) and the physical structure (storage structure, access methods, multidimensional indexing) of the spatio-temporal data model are also described in the book.

This book is not an introduction to object-orientation. There already exists a vast literature regarding this subject. The motivation for this book is the relative absence of research in the field of integrating spatio-temporal data modelling, object-orientation and GIS. Therefore, the book offers a practical guide to object-oriented modelling constructs and demonstrates the feasibility of applying them to the problem of modelling spatio-temporal data. It encourages readers to apply and explore object-oriented analysis and design methods by presenting a variety of practical examples in the application area of political boundary record maintenance (historical data). A prototype implementation of the spatio-temporal data model into the Smallworld GIS illustrates the results of the processes involved in boundary making of public boundaries in England.

This book is a revised and summarised version of my doctoral thesis submitted to the Department of Geography at the University of Edinburgh. It is intended for readers from multi-disciplinary research areas. It is required reading for all those interested in the role of object-orientation for integrating space and time in GIS. The book is essential reading for students, scientists and researchers in the fields of Geography, GIS, Cartography, and Databases.

It begins with a synopsis of space and time concepts, their application in GIS, and the functional requirements to develop a temporal GIS. A variety of applications and the characteristics of their spatio-temporal data are described. A literature review is included on Time Geography, temporal GIS and the object-oriented paradigm (programming languages, analysis and design methods, databases). Future research directions are outlined in spatio-temporal data modelling, version management, and applications with complex and interrelated objects and spatio-temporal data. This book provides an objective source of references for addressing conceptual, methodological, and technical issues for handling spatio-temporal data in GIS.

The object-oriented analysis and design method developed by Booch is used in this book. This method has been the most widely accepted in the database research and development communities. The synergy of Time Geography, object-orientation and GIS is illustrated by a prototype implementation. Finally, the last chapter discusses issues that are emerging as important areas of technological innovations in GIS. Knowledge Discovery in Databases (KDD), geographic visualisation (GVis), and ubiquitous computing reinforce the use of the object-oriented methodology in GIS.

I have attempted to provide the reader with an understanding of the issues in the object-oriented design of a temporal GIS and have taken a position on most of them: comments, corrections, and suggestions welcome.

Acknowledgements

I am thankful to several people who encouraged me in one way or another to realise the original thesis which led me to write this book. First of all, I would like to thank my supervisors, Professor Richard G. Healey and Professor Michael F. Worboys, for their unceasing support and valuable advice during my thesis research. My thanks also go to Les J. Rackham and Alan Hardiman of the Ordnance Survey, who have significantly contributed in expanding my knowledge on the evolution of public boundaries in Great Britain.

The Department of Geography at the University of Edinburgh has provided a stimulating environment in which to work and my thanks go to those who helped to make it that way. In particular, I would like to thank Steve Dowers and Chris Place for their support and expertise on computing facilities in the Department, and I would like to thank Penny Leg for making my research meetings with Richard possible.

I was fortunate to find many colleagues in Edinburgh who were willing to listen to my ideas; in particular Bhaskar Ramachandran, Trevor Rotzien, Charles Stewart, Jonathan Makin and Christoph Corves who provided thought-provoking discussions and suggestions. I would like to thank Marianne Broadgate, Vicen Carrio-Lluesma, Juan Suarez, Manoel Claudio da Silva, Marcelo Vieira, Gwo-Jinn Hwang, Xuan Zhu and Bod Hodgard for their friendship and endless support during my years in Edinburgh.

The Research monographs in Geographical Information Systems series has provided me with the possibility of publishing my thesis, and I would like to thank Professor Peter Fisher and Doctor Jonathan Raper for the opportunity of transforming my thesis into this book.

I am grateful for the financial support and encouragement I have received from my family. Distance could never deter them from being a continuous source of love and reassurance. At the risk of under-expressing my gratitude, I would like to thank my husband Victor for his unconditional optimism, continuous source of love and support. This book is dedicated to them.

CHAPTER ONE

Introduction

Geographical information science has recently emerged as a distinct interdisciplinary knowledge field involving many diverse areas such as geography, cartography, engineering and computer science. In this field, geographic information systems (GIS) have been used for analysing spatio-temporal data sets pertaining to social, environmental and economic studies. This has led to the integration of a variety of socio-economic and environmental models with GIS. Examples include the innovative GIS-based monitoring model developed by Blom and Löytönen (1993) to monitor current epidemics in Finland, including HIV. This model integrates spatial diffusion, spatial interaction and environmental modelling into a GIS-based model for monitoring the passing of infectious diseases between individuals. The goal of this model is to provide disease-specific forecasts for the future course of an epidemic.

The European Groundwater Project (Thewessen, Van de Velde and Verlouw, 1992) is one example of the integration of existing non-spatial simulation models with spatial data sets. The result is the design of a GIS-based environmental model that provides rapid and coherent access to the most significant causes and effects of groundwater contamination. Physical and chemical models have been integrated into the GIS-based model so it can identify serious threats to the quality and quantity of groundwater resources in the European Union.

The integration of the CLUE model (conversion of land use and its effects) with a GIS is an example of a dynamic, multi-scale, land use change model developed to explore the complexity of the interactions between socio-economic and biophysical factors in land use changes. It was applied to data from China, Ecuador and Costa Rica (Verburg et al., 1997). The results indicate the importance of understanding the dynamics of land use within a multi-scale scenario. Implementation of such a model was essential to explore the spatio-temporal patterns of land use change under different scenarios of population growth and food demand.

Researchers and developers are continually uncovering different uses for GIS-based models in non-traditional applications. Burrough and Frank (1995) draw attention to the diversity of ways of perceiving the same knowledge domain, and consequently the proliferation of many models for handling the knowledge domain at different levels of complexity as well as aggregation in GIS. The study of common concepts and principles among these models is essential when formulating design criteria and strategies to support and advise users on how to integrate them in a GIS. An array of possibilities and new perspectives are expected to arise on how this could be achieved. This book proposes the *object-oriented paradigm* as a common framework to handle the complexity of semantics of spatio-temporal data defined within a knowledge domain.

1.1 OBJECT-ORIENTED ANALYSIS AND DESIGN

Object orientation in modelling spatio-temporal data has been widely recognised as a powerful tool that captures far more of the meaning of concepts within a problem domain (Rojas-Vega and Kemp, 1994; Milne, Milton and Smith, 1993; Worboys, Hearnshaw and Maguire, 1990). It enhances the level of abstraction in a way close to our perception of the real world, offering a mechanism for expressing our under-standing of the knowledge domain. Jackson (1994) advocates the use of object-oriented modelling in regional science as a common framework for integrating different semantics defined within social models. Object orientation is presented as a systematic approach to modelling the conceptual descriptors of complex socio-economic models. It provides a way to formalise the handling of problems that need to be solved by the combined efforts of several people.

Bian (1997) has used the object-oriented paradigm to extend a two-dimensional static growth model into a three-dimensional dynamic framework. The aim was to study individual fish behaviour in an aquatic environment. In his object-oriented salmon growth system, the movement of individual salmon in a three-dimensional space was incorporated with the growth model to simulate the behaviour of salmon in selecting their habitat and their consequent growth. A number of simulations were run with five to ten adult salmon at a time for a period of several days.

However, the complexity of integrating object-oriented and geographic concepts into a spatio-temporal data model is an interesting challenge in its conception and its implementation. Choosing an object-oriented method is a laborious task. Object-oriented methods have been introduced into several distinct structures and repres-entations, with over 50 published suggestions. 'They range from the complex and difficult notations of OMT, Ptech and Shlaer/Mellor to the simpler ones of CRC and Coad/Yourdon, from an emphasis on process to an emphasis on representation and from language dependence to the giddiest heights of abstraction. . . . None of these methods is complete in the sense that all issues of the software development life cycle are addressed or that every conceivable system can be easily described' (Graham, 1994, p. 287).

This book summarises a significant amount of research carried out in object orientation. Many of the concepts and implementations developed in this area are discussed and brought together within the context of GIS. The objective is to provide readers with a solid understanding of the object-oriented paradigm for designing a spatio-temporal data model.

1.2 SPATIO-TEMPORAL DATA IN GIS

Representing spatial data in a GIS has been achieved by defining entities in geometric space in an explicit manner (vector representation) or an implicit manner (raster representation); see Burrough (1986). In the vector representation, three main geometric elements are used: points, lines and polygons, which are sets of vectors with interconnected coordinates linked to given attributes. The relationship among elements is represented by the connectivity of a set of vectors at the time of their storage into a GIS. For example, a set of lines is represented by starting and ending points, and some form of connectivity (straight line, curve, etc.). In a raster representation, entities are sets of cells located by their corresponding coordinates. In this case each cell is linked to an attribute value. The location of each cell is used to determine the adjacency relationship between entities.

As Dutton (1987) points out, the debate on vector versus raster representations is nearly as old as the concept of GIS. Both representations of geographic space have been regarded as valid data models. Besides, data transformation algorithms to convert from one spatial representation to another have been developed, and the choice between them is taken by the user who selects the representation that is most efficient for implementing a particular application in a GIS. Consequently, GIS has fully developed into information systems that are characterised by capabilities for representing, querying and manipulating entities in space. Over the past decade, expectations about exploring spatio-temporal data in GIS have raised interest in a wider range of capabilities. Some of these capabilities can be described as update procedures that are coherent with previous stored data, version management mechanisms to track the lineage of data, and analytical tools to recognise patterns of change through time as well as to predict future changes.

Representing spatio-temporal data in a GIS has been regarded as implementing an additional dimension in a former spatial representation (vector or raster). The primary objective for most of the spatio-temporal representations is summed up in the idea *organising space over time*. A geographic space is organised into partitions (layers) and the entities that inhabit this space are embedded in these partitions. In fact, a partition serves as a skeleton for representing several entities located in the geographic space at a particular point in time. This is a *region-to-entity* representation: first choose a region of a geographic space, then identify and locate the entities that inhabit that region according to how alike they are or how they are composed. Space and time dimensions are incorporated by determining their singularity through their contents; for example, space by attributes and shapes of the elements (points, polygons, lines, grid cells) and time by succession of happenings (events, actions, change, motion) on these elements. So far, this approach has been used in GIS by making spatially depicted classifications grouped into layers or sets of themes (e.g. geology, hydrology and land cover) between points or periods of

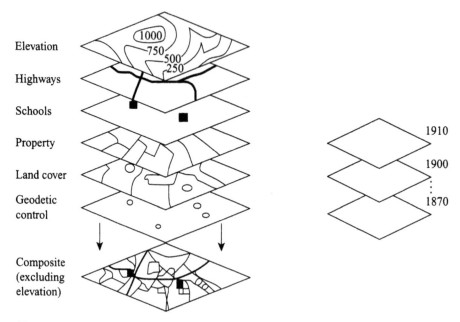

Figure 1.1 Spatio-temporal layers as the main representation being used in GIS (Reprinted with permission from Laurini and Thompson 1992, Academic Press Ltd)

time. In other words, geographic space is grouped along the spatial dimension after some sort of categorisation, and time is grouped along the time dimension after some sort of periodisation. Constituting history is explained based on similarity or dissimilarity between aggregations (layers) at different points of time (Figure 1.1).

Although this four-dimensional representation is sufficiently homogeneous for capturing and storing spatio-temporal data in GIS, it does not provide a unified representation of the real-world. We are dealing with geographic space: a space that reflects our knowledge of the environment where time exerts its influence on place in terms of human tasks and lived experiences. If we could decide, once and for all, which real-world phenomena should be represented as entities, relations or attributes in a geographic layer, our modelling task would be extremely simplified. In fact, what we need is to understand the nature of time itself with respect to the real-world phenomenon that we are trying to represent in a GIS. In order to accomplish that, the emphasis must shift from organising space over time to *representing a real-world phenomenon in space and time*.

This representation gives us an entirely different perspective to how we handle spatio-temporal data in GIS. It attempts to capture the complexity of space and time at the level of an indivisible unit – the entity. Instead of creating layers or time periods, this representation deals with elements' coexistence, connection or togetherness. We are distinguishing two important concepts that are often regarded as interchangeable, an 'entity' and an 'entity embedded in space'. This distinction would be unnecessary if we could always define the precise location of entities and their corresponding classified layers or time periods. In fact, we are confronted with a rather different reality. Most likely is that we may be uncertain of their location and how they change or move in a dynamic way. Moreover, we may know the location of an entity in a geographic space but we are uncertain of how to classify it. The notion of having an entity unconstrained by its surroundings in space and time allows us to examine how a real-world phenomenon is represented independently of how geographic space is organised at a particular time.

This is a *space-time entity* representation: first identify the entities, and second ensure that based on these entities a geographic space can be created. An important characteristic of this representation is the ability to create the geographic space based on a specific task to be solved or a particular knowledge about the real-world at a particular point in time. Depending on the specific task to be solved or the human ability to see the world at a particular point in time, certain real-world phenomena may be represented as entities in a geographic space, and others become the relations we are interested in modelling. For other tasks or different perspectives in the world, these roles may change. Therefore, modelling spatio-temporal data in GIS becomes an exercise of understanding not only the similarities and dissimilarities between regions of geographic space, but also the coexistence (connection or togetherness) relationships between the entities that inhabit these regions.

A reliable space-time entity representation is needed when designing a spatio-temporal data model in GIS. As Peuquet points out, a variety of approaches for studying space-time phenomena has evolved in social, geographical and physical studies. 'Andrew Clarks's early work on historical geography demonstrated that changing spatial patterns could be studied as "geographical change" (Clark, 1959, 1962). Cliff and Ord (1981) later examined change through time by scanning a sequence of maps, searching for systematic autocorrelation structures in space-time in order to specify "active" and "interactive" processes. Perhaps the best-known efforts within the field of geography that made explicit use of time as a variable in the study of spatial processes are Hägerstrand's models of diffusion and time geography' (Peuquet, 1994, p. 441).

1.3 TIME GEOGRAPHY

Torsten Hägerstrand, a Swedish geographer, unfolded the Time Geography approach in the early 1960s. He examined space and time within a general equilibrium framework, in which it is assumed that every entity performs multiple roles; it is also implicitly admitted that location in space cannot effectively be separated from the flow of time. In this framework, an entity follows a space-time path, starting at the point of birth and ending at the point of death. Such a path can be depicted over space and time by collapsing both spatial and temporal dimensions into a space-time path. Time and space are seen as inseparable.

Time Geography has provided a foundation for recognising paths of entities through space and time and for uncovering potential spatio-temporal relationships among them. Moreover, its application in various areas has produced the concept of a 'continuous path' to represent the experience occurring during the lifespan of an entity. This experience is in fact conceptualised as a succession of changes of locations and events over a space-time path. Most of the applications using Time Geography have been devoted to modelling individual activity paths within a period of time, analysing the pattern of activities for any individual path, as well as simulating individual activity paths.

This book proposes a new means for applying the time geography approach. Its goal is to employ the concept of a space-time path developed in time geography for representing spatio-temporal data within a spatio-temporal data model. The time geography framework introduces a robust space-time entity representation for conceiving a spatio-temporal data model. In this case, time geography plays an important role as a *modelling tool* for representing the passage of time and the mechanisms of change within a spatio-temporal data model. This approach for dealing with time and space within a GIS has not been explored up to now, and the book attempts to demonstrate a new and more encompassing perspective for integrating space and time domains within a GIS. The time geographic spatio-temporal data model proposed here will be known throughout the book as the spatio-temporal data model (STDM).

1.4   THE SPATIO-TEMPORAL DATA MODEL

The STDM proposed in this book involves conceptual and implementation considerations that present a variety of semantic and structural aspects to be dealt with. The range of aspects can vary from addressing the complex and subtle spatio-temporal semantics of a real-world phenomenon to the development required for the logical components (schema evolution, query language syntax) and the physical structure (storage structure, access methods, query optimisation) of the system.

Therefore, the analysis and design of such a spatio-temporal data model can be fraught with a whole assortment of problems. These are essentially related to our understanding of the knowledge domain, the modelling constructs, and the mapping between the model and its implementation in a GIS. The use of object orientation is required in order to obtain the space-time entity representation for the spatio-temporal data model and the design tool for implementing this model into a GIS. Object-oriented methods offer a concise methodology that allows us to focus our attention on the conceptual aspects of the system, and to concentrate on the details of the design without being overwhelmed (Rubenstein and Hersh, 1984).

The book also encourages readers to apply and explore the STDM by presenting a practical application of political boundary record maintenance (historical data). The chosen application deals with the evolution of public boundaries in England. The Ordnance Survey is the national mapping agency for Great Britain which 'has had a statutory requirement to ascertain, mere and record public boundaries since 1841. As a result, it has become the main depository for, and authority on, public boundaries in Great Britain' (Rackham, 1987, p. 6). On 1 April 1991 the Ordnance Survey created a spatial data set at 1 : 10 000 scale containing the digital outlines of the public boundaries in England. In order to support this data set, the Boundary-

Line system has been defined; it produces snapshots showing the location of public boundaries at specific dates. This pioneering initiative has been influential in consolidating the perspective of this research towards the design of a spatio-temporal data model that can contribute in a number of ways to the development of the Boundary-Line data management system used by the Ordnance Survey.

The implementation of the STDM in Smallworld GIS is undertaken as a 'proof-of-concept'. Implementing the STDM has been the means by which the ideas developed in the model could be empirically tested. This book describes the implementation aspects of STDM, highlighting the challenges for geographical information science.

1.5 AIMS OF THIS RESEARCH

This book introduces a spatio-temporal data model which integrates space and time domains in a GIS context, based on the concepts developed in the Time Geography and object-oriented approaches. The research had five aims:

1 Define the space-time entity representation as a new means of characterising spatio-temporal data in GIS.

2 Provide a deeper understanding of the meaning of space-time paths and use this to identify a suitable role for dealing with the passage of time and the mechanisms of change within a spatio-temporal data model in GIS.

3 Converge both approaches: Time Geography and object orientation, by associating space-time paths of a time geographic framework with the modelling constructs of an object-oriented method.

4 Contribute to the development of the Boundary-Line data management system of the Ordnance Survey by providing a different perspective about spatio-temporal data modelling in GIS.

5 Undertake the implementation of the spatio-temporal data model into a GIS system as 'proof-of-concept'.

1.6 ORGANISATION OF THIS BOOK

Chapter 2 introduces the main concepts involved in the Time Geography approach that have been used for developing the spatio-temporal data model. The feasibility of incorporating this approach into a GIS is discussed on the basis of the previous implementation efforts that have been found in the literature. Chapter 3 provides a historical background to object orientation by summarising the efforts in the areas of object-oriented methods, temporal databases and version management approaches. The object-oriented analysis design proposed by Booch (1986, 1991, 1994) is presented as the best-worked-out notation and technique for integrating the time geography framework into our spatio-temporal data model.

Chapter 4 presents the spatio-temporal data model based on time geography and object orientation concepts previously described in Chapters 2 and 3. Chapter 5 considers how to apply the spatio-temporal data model to boundary-making for public boundaries in England. A comprehensive set of diagrams demonstrates the important aspects of the spatio-temporal data model. Chapter 6 presents the results from implementing the spatio-temporal data model. A prototype implementation illustrates the working of the spatio-temporal data model. Chapter 7 discusses the emerging technologies relevant to geographical information sciences, and provides future research ideas for possible advances in spatio-temporal data modelling.

CHAPTER TWO

Concepts of space and time

Time and the way it is handled has a lot to do with structuring space.

E. Hall, *The Hidden Dimension*

This chapter is a brief guide to some concepts in the literature on temporal GIS. The Time Geography approach is introduced as a modelling tool for representing the passage of time and the mechanisms of change within a GIS. The main concepts involved in Time Geography which have been used for developing our spatio-temporal data model are described in this chapter. The feasibility of incorporating this approach into a GIS is discussed on the basis of previous implementation attempts.

2.1 THE SPACE-DOMINANT VIEW

Although time and space are concepts inherently related, we encounter difficulties in thinking and hypothesising about them in equal terms. Langran (1992a) has coined the term 'dimensional dominance' to illustrate how our discernment of space and time in GIS has been influenced by space-dominant or time-dominant representations. The space-dominant representations focus on the spatial arrangement of entities based on the geometric and thematic properties of those entities. In other words, attention is given to the spatial arrangement as an ensemble of phenomena in a geographic space and not so much to a phenomenon itself. The space arrangement is perceived as a *layer* that can combine a variety of themes and efficiently be used for storing and processing spatial data. Fisher (1997, p. 301) points out: 'The idea that the world can be broken up into its constituent themes (layers) which can be treated independently of each other is endemic. . . . It is seen as having the advantage of simplifying a complex world'.

The concept used here is of *absolute space*, which considers space as infinite, homogeneous and isotropic, with an existence fully independent of any entity it

Table 2.1   Main characteristics of the space-dominant view.

- Space is viewed as a container
- Elements only exist when associated to a layer or theme
- Applied primarily in traditional mapping
- Layer-based raster and vector models
- Each layer is associated to a period or point in time
- Change- or update-based scenario
- Analysis based on similarity or dissimilarity between aggregations (layers) at different points of time

---

might contain. Time is implicitly incorporated into the spatial arrangement every time some sort of change occurs. As a result, a snapshot of a layer is created every time an update occurs. A sequence of snapshots describes the passage of time. However, it is not possible to know how an updated layer might affect other associated layers of the same geographic space. Today GIS products support some sort of spatial-dominant representation, i.e. layer-based raster or vector models. These models present spatially depicted classifications grouped into layers or sets of themes (e.g. geology, hydrology and land cover) between points or periods of time. In other words, geographic space is grouped along the spatial dimension after some sort of categorisation, and time is grouped along the time dimension after some sort of periodisation. Constituting history is explained based on similarity or dissimilarity between aggregations (layers) at different points of time (Figure 1.1). Topographic mapping, navigational charting, utility mapping and cadastral mapping are some examples of space-dominant domains.

Peuquet (1994) points out that absolute space is objective since it give us an immutable structure that is rigid, purely geometric and serves as the framework in which entities may or may not change (change- or update-based scenario). This is probably the reason why most GIS products have adopted the space-dominant view within their data models (Table 2.1). Clifford and Ariav (1986) describe various examples of modelling change in the space-dominant domains. Most of the examples extend the relational database model by creating new versions of tables, tuples or attributes every time a change occurs. Their main conclusion was that change is best incorporated as a component of the database at the attribute level, rather than at the tuple or table level. The main reason was that by associating a time stamp with each attribute, the user has more control over the semantics of the data, and more flexibility in the kind of queries that can be posed. They also argue that time stamping attributes provide database management systems (DBMS) with greater flexibility in both storage and query evaluation strategies.

Langran (1989) also reviews temporal GIS research on the basis of dimensional dominance and concludes that attribute versioning is a hybrid organisation which offers the most adequate approach for GIS applications presenting spatial dominance. Although time is generally perceived as continuous, the preference for a discrete time representation stands out in space-dominant domains. Time is treated as a discrete subset of the real numbers ordered linearly. Therefore, changes are supposed to take place a finite number of times so that each change produces a sequence of historical states indexed by time.

2.2 THE TIME-DOMINANT VIEW

When time takes part explicitly in a representation, either with or without reference to space, the time dominance is generated and an absolute view of time is used within a model. In this case the chosen concept is *absolute time* as a fourth dimension, a time line marked out with intervals, and along which events, observations or actions can be located. This representation is effective in domains where the accuracy of the temporal information makes it possible to date or order events, observations or actions. It presents a time structure (temporal logic), and the statements about events, observations or actions are either true or false at various points in the time structure.

Al-Taha and Barrera (1990) present a first attempt to classify time-dominant representations into three categories:

- *Interval-based models* where temporality is specified using regular or irregular intervals (Allen, 1983). The representation deals with identifying temporal intervals by defining relationships between these intervals in a hierarchical manner. In this case, a specific date is not necessary; relationships between two intervals are instead defined in the model. The relationships are before, equal, meets, overlaps, during, starts and finishes. Allen (1983) asserts that with these relationships one can express any permanent relationship between two intervals.

- *Point-based models* where temporality is specified using explicit occurrences of an event, observation or action (Dean and McDermott, 1987). These models are usually implemented as time maps. A time map is a graph whose nodes refer to points of time that correspond to the beginning and ending of an event, observation or action. The edges represent the relationship between events, observations or actions.

- *Mixed models* where temporality is specified using an interval-based model combined with a point-based model (Shoham and Goyal, 1988).

These models have not been implemented in GIS, where temporal capabilities are not yet fully developed. But there is a need for handling large amounts of data that involve time. Archaeological data and geological data are two examples where precise dates for events are not known but the relative order can be deduced. On the other hand, inventory data and environmental data are examples of time series where the precise date of each observation on a particular variable is known, and the sequence of observations provides the occurrence of a real-world phenomenon (Table 2.2).

Table 2.2   Main characteristics of the time-dominant view.

- Time is viewed as a time line
- Events, observations or actions are associated to a time line
- Applied in archaeology, geology and environmental sciences
- Interval, point and mixed models
- Space is not an entity in itself
- Event-based scenario
- Analysis based on the lineage of events, observations or actions

Nevertheless, the semantics of time have been incorporated in GIS using different approaches. They can be distinguished according to the assumption of time as a parameter or dimension (Effenberg, 1992). In the parameter approach, time is employed as a control argument within the system while possible effects over other variables are investigated. This approach is largely employed in simulation modelling in GIS. On the other hand, the dimensional approach has introduced a dynamic construct in GIS. The time dimension is implemented as a user-defined data type. For example, Illustra has implemented a time series data type that consists of information on the calendar observed by the time series, the starting time of the time series and the stride between observations, e.g. daily or monthly (Stonebraker and Moore, 1996).

2.3   THE ABSOLUTE SPACE-TIME VIEW

Both space- and time-dominant views have influenced research outcomes since the early 1980s. Armstrong (1988) has defined eight possible combinations of changes or updates which can occur in vector-based models. For each possible update procedure, a change is associated with the geometry, topology and thematic properties of an entity in space. Kucera (1996) has also advocated the need for developing data-driven update procedures in GIS, procedures based on where and when the change occurs.

TEMPEST (Temporal Geographic Information System), proposed by Peuquet (1994), is the first effort towards the integration of space- and time-dominant views in GIS. 'Location in time becomes the primary organisational basis for recording change. The sequence of events through time, representing the spatio-temporal manifestation of some process, is noted via a time-line; i.e., a line through the single dimension of time instead of a two-dimensional surface over space [see Figure 2.1]. . . . Such a time-line, then, represents an ordered progression through time of known changes from some known starting date or moment to some known, later, point in time' (Peuquet and Wentz, 1994, p. 495).

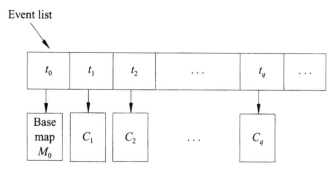

Figure 2.1 The representation of change organised as a function of time in the TEMPEST prototype

2.4 THE RELATIVE SPACE VIEW

For most of our spatio-temporal analysis, the relative view of space is of the most fundamental importance. The concept of *relative space* is more general and empirically more useful than the concept of absolute space. Jammer (1969, p. 23) defines relative space as 'an ordering relation that holds between bodies and determines their relative positions . . . a system of interconnected relations'. The profound implication is that any relation defined on a set of entities creates space. In other words, defining a relation automatically defines a space. Harvey (1969) provides an excellent review of the two perspectives, absolute space and relative space. The concept of absolute space overemphasises the absolute location of entities within a spatial representation. In contrast, relative space focuses on the relative location among entities. The relativistic point of view is usually associated with studies of forms, patterns, functions, rates and diffusion.

The study of gradual changes of topological relationships has recently emerged as a requirement in formalising a spatio-temporal representation in a GIS. Egenhofer and Al-Taha (1992) have investigated gradual changes in the location of an entity, such as translation, scaling and rotation, by formalising them using eight binary topological relationships for two spatial regions. The eight binary topological relations are depicted in the closest topological relationship graph showing the links between gradual changes in topology. Each gradual change allows many possible scenarios; one of them is illustrated in Figure 2.2.

2.5 THE RELATIVE TIME VIEW

Another important concept is *relative time* – time measured in relation to something, not constrained to a single dimensional axis. Cyclical time – the repeating of a day, week or year – is an example of relative time. In absolute time 13 August 1998 cannot be repeated. But in relative time, Thursdays keep returning. Most questions about change will be understood from this perspective (Ornstein, 1969).

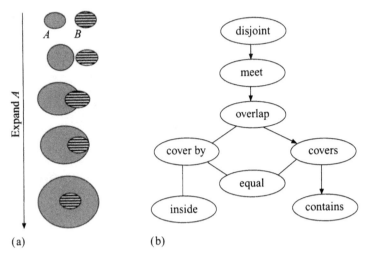

Figure 2.2   Example of a sequence of gradual topological changes between two entities

Relative time is subjective since it assumes a flexible structure that is more topological in the sense that is defined in terms of relationships between events. Frank (1994) suggested an ordinal model of time in which a chronological order is defined among events of a time line rather than attaching precise dates for these events. Some examples are the qualitative ordinal information about events that is typically encountered in archaeology and urban development. The precise dates for events are not known but the relative order of events can be deduced from observations.

2.6   THE RELATIVE SPACE-TIME VIEW

The relative space-time view embraces human activity over the real-world that results from studying processes within an application domain: 'A process study seeks to identify the rules which govern spatio-temporal sequences, in such a form that the rules are interpretable in terms of the results of the sequence, in terms of the exogenous variables which influence the sequence, and in terms of the mechanisms by which exogenous and endogenous influences give rise to the results which the sequence itself records' (*Dictionary of Human Geography*, 1994, p. 478). Table 2.3 summarises the main characteristics encountered in the relative space-time perspective.

Gatrell (1983) provides several examples of constructing space-time maps based on proximity relations among entities. The approach given is the multidimensional scaling (MDS) algorithm, in which relations are defined by numerical values in a matrix representing perceived distances between entities (main cities in New Zealand) or their rank orders over time. Figure 2.3 shows the result of an MDS algorithm for representing New Zealand in space and time.

Table 2.3 Main characteristics of the relative space-time view.

- Space and time are viewed as coexistence (connection or togetherness) relationships between changes and events
- Neither space nor time exists independently
- Applied in studies of forms, patterns, functions, rates and diffusion
- Topological models
- May involve non-Euclidean space or linear time
- Process-based scenario
- Analysis based on a process study

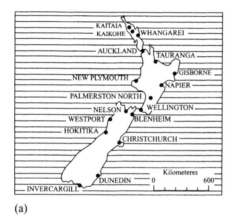

(a)

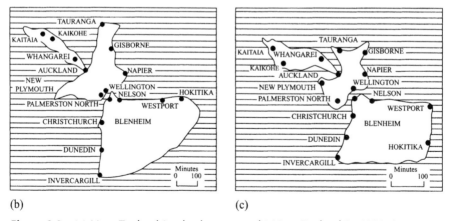

(b)   (c)

Figure 2.3 (a) New Zealand in absolute space (b) New Zealand in 1953 time space (c) New Zealand in 1970 time space. (Reprinted by permission of Oxford Press. *Source* Gatrell 1983, p. 111)

2.7   CHOOSING THE VIEW FOR A GIS

Harvey argues that we 'have (frequently) assumed a particular spatial language (i.e. absolute view or relative view) to be appropriate without examining the rationale for such a choice' (Harvey, 1969, p. 161). After all, we should not discriminate one over another. They are complementary. As Peuquet (1994) points out, the absolute view requires some sort of measurements referenced to a constant base, implying non-judgmental observation. The relative view, on the other hand, involves interpretation of processes and the flux of changing patterns within a knowledge domain.

However, a question still remains about integrating absolute and relative views. How can we have both perspectives placed in the same representation? Perhaps the answer lies in time geography: 'Owing to the circumstances under which . . . [Time Geography] has been developed, its contents, and its applications to date, there is a great danger that Hägerstrand's time geographic framework will be mistakenly construed as nothing more than a planning tool. On the contrary, . . . the potential usefulness of the framework . . . is of much greater range' (Pred, 1977, p. 213). Consequently, this book proposes that the concepts of Time Geography should be exploited by GIS to capture the absolute space-time view as well as the relative space-time view. The next section provides an overview of Time Geography.

2.8   TIME GEOGRAPHY

The pioneering work of Torsten Hägerstrand during the 1960s unfolded the Time Geography research that emerged from the Royal University of Lund in Sweden. The concepts developed in time geography have been mostly consolidated in the work of Hägerstrand and his students and collaborators Lenntorp, Mårtensson and Carlstein. Pred (1977) provides an overview of the main uses of Time Geography in several domains, among them domains concerned with regional development policies, nationwide physical planning, and urbanisation and settlement policies. The Swedish government has implemented many of the applications of Time Geography in order to provide adequate job-market opportunities, and a satisfactory level of social and cultural services. Some examples are the accessibility simulation of daily individual activities in urban environments and regions (Lenntorp, 1978), comparative studies of living conditions in different populated regions (Mårtensson, 1978), and analysis of various activities in the quaternary sector, mainly concerned with employment distribution (Olander and Carlstein, 1978).

Space and time in Time Geography are considered as orthogonal dimensions that become fused into a space-time path representing the trajectory for the lifespan of an entity (Figure 2.4). For simplification, as suggested by Hägerstrand, the representation of space is along only one dimension to maintain the clarity of the proposed representation framework and to give a better visualisation of the evolution of public boundaries. The same simplification has been adopted for the time dimension.

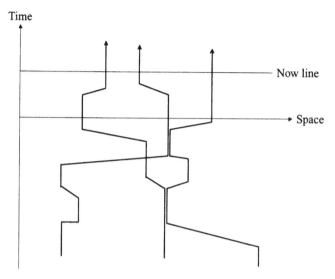

Figure 2.4 Space-time paths of three entities

Time is viewed as the dimensional axis that orders events, separates causes from effects, and synchronises and integrates human activity (absolute time representation). Space is viewed as the dimensional axis that represents the changes in the location of an entity in space (absolute spatial representation). Space and time are joined in a single space-time path. Space and time are considered inseparable within a path, and it is the timing component which gives structure to space and thus evokes the notion of *place*. Place is 'a pause in movement' (Parkes and Thrift, 1980, p. 120).

The Time Geography approach is an effort to capture the complexity of space-time interaction at the scale of the smallest indivisible unit in space and time, i.e. the space-time path (relative space-time view). However, by having space and time as orthogonal dimensions, Time Geography requires the absolute view of space and time. An absolute location in the space axis and an absolute location in the time axis are needed to define a place on a space-time path. The absolute location in the space axis may be determined with reference to a coordinate system, whereas the absolute location in the time axis may refer to location in time derived from a clock or calendar. A space-time path provides the starting time, the duration, the frequency, the sequential order, as well as the relative location of events and changes that have occurred in a lifespan of an entity.

The continuous line of a space-time path is the relative space-time representation of the lifespan of an entity. It is a descriptive form to reveal the interdependence and relationships between events and changes (Lenntorp, 1976). 'Space and time are to be jointly treated ... because when events are seen located together in a block of space-time [paths], they inevitably expose relations which cannot be traced if those events are bunched into classes and drawn out of their place in the block, i.e., conventionally analyzed' (Pred, 1977, p. 210).

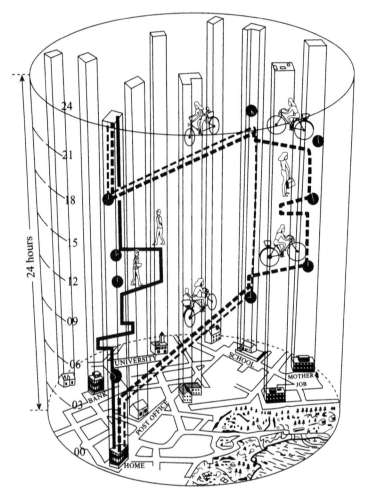

Figure 2.5   An example of potential path areas. (Reprinted with permission from Parkes and Thrift, 1980, John Wiley & Sons Ltd)

The sense of past, present and future depends on where the observer is placed in the space-time path. The observer has the awareness of coexistence (connection or togetherness) relationships between space and time at every place located in a space-time path (Hägerstrand, 1975). The motion of an observer can be limited to a set of circumstances described by the constraints which have been defined for a space-time path. Called the potential path areas (Figure 2.5), this set is represented as a prism in Time Geography. It comprises space-time positions for which the possibility of being included in the observer's trajectory is greater than zero (Lenntorp, 1976). A general procedure cannot be developed for deriving or calculating potential path areas from empirical data. Each knowledge domain has to be analysed in order to generate its actual potential path areas.

2.9 TIME GEOGRAPHY AND GIS

Although Time Geography is an effective approach to dealing with space and time in an integrated manner within a GIS, it has so far been neglected. After analysing the feasibility of handling space-time concepts of Time Geography within a GIS, Miller affirms that 'Geographic Information Systems, through their ability to manipulate and analyse spatial data, can allow more widespread use of the space-time perspective [of Time Geography] in spatial modelling and analysis' (1991, p. 300).

Few examples are available for illustrating the attempts at applying the Time Geography approach within a GIS. Miller (1991) has generated potential path areas (PPAs) for a transportation network application on the basis of arcs in the network that are feasible to travel. The mainframe version 5.0 of ARC/INFO was used for the implementation in order to handle a set of nodes and arcs, keep records of locations within the system, and handle numerous travel times at both nodes and arcs in the network. Although 'ARC/INFO NETWORK can meet the requirements for standard GIS applications; it is inefficient and unwieldy in meeting the more specialised needs of the network PPA procedure. Whereas ARC/INFO is certainly not representative of all GIS software, it does provide a benchmark which indicates the problems encountered by analysts who wish to use GIS technology in more specialized research and modelling' (Miller, 1991, p. 299).

Miller also points out the main requirements in applying Time Geography in a GIS:

- The Time Geography approach requires data at a detailed spatial scale in order to obtain an effective analysis in a GIS.

- The GIS must be able to address the behavioural aspects of data to generate more realistic operational PPAs.

- The favoured GIS to implement a time geographic framework must be able to store and manipulate topological relationships to avoid adding unnecessary complexity to the framework.

- The derivation and manipulation of PPAs in a GIS might be accomplished by developing the framework to support space and time constraints. A modular structure with inflexibility of key commands and procedures can render a GIS unable to derive the desired space-time prism framework.

Another example is the application of Time Geography for simulating an individual's daily shopping behaviour within a GIS (Makin, 1992). The results show how time and space constraints on people's shopping movements affect shops' potential earnings and profits. Makin explores the potential of using a GIS to structure spatial relationships according to which routes are accessible to each other, and where the buildings are located on the route network.

He also points out the potential benefits of having implemented his Time Geography model into Smallworld GIS for simulating the behaviour of people's movements:

- The data items are not generalised or aggregated within the GIS.

- The entities are allowed to move and interact with their constraints in space and time in a way that long-term behavioural patterns can be analysed.

- The model is expressed in terms that do not require abstraction into mathematics.

- The whole system can be organised in a modular fashion in which subsystems are created for reducing the complexity of the model by minimising the amount of data and the number of interactions.

These two examples illustrate the potential perspectives of applying Time Geography in a GIS. And Time Geography could be used to formulate space-time semantic abstractions in the design of a spatio-temporal model based on the object-oriented approach – something not thoroughly explored until now.

2.10   MAIN ELEMENTS OF A SPACE-TIME PATH

Space-time paths can be defined as image schemata that directly depict the lifespan of each singular entity in a knowledge domain. The space-time path is based on the fact that an observer can move anywhere along the space-time path, there is a starting location (in space and time), a direction (motion, change), and a sequence of continuous locations (in space and time) that the observer goes across in following the path. The space-time path embodies the structures of identity, location (in space and time) and direction that are the basic abstract metaphors for modelling spatio-temporal data in GIS.

Langran (1992b) asserts there are at least three sorts of spatio-temporal data in GIS: states, events and evidence. In extending this classification, Kraak and MacEachren (1994) point out a further differentiation between events and episodes. In essence, a *state* represents a version of what we know about an entity in a given moment. States can consist of different versions of an individual entity (the changes of an individual political boundary marking the variation on the distribution of a territory and its sovereignty) or an ensemble of entities (the changes in ecosystem conditions produced by reductions in atmospheric emissions of pollutants). An *event* is the moment in time an occurrence takes place. Events cause one state to change to another (e.g. a cadastral survey may take place due to changes on properties' boundaries). Events are also part of a process of change caused by action and reaction as well as the synthesis of both (e.g. the process of energy propagation and ozone hole formation). An *episode* is defined as the length of time during which change occurs, a state exists, or an event lasts. And finally, a piece of *evidence* is the datum describing the source of state and event data. No evidence should ever be stored in a GIS without referencing its source document, survey or update procedure.

This book proposes the space-time paths as ideal image schemata for representing and organising the spatio-temporal data categories in GIS (state, event, episode and evidence). The aim is a better understanding of the defined categories by distinguishing the three main structures of a space-time path: identity, location and

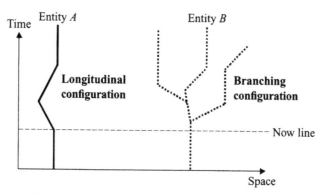

Figure 2.6 Possible configurations for a space-time path

direction. Any knowledge domain contains a population made up of entities which are represented in the Time Geography as point objects. A singular space-time path describes a lifespan trajectory through time over space, from the point when and where the entity comes into being, to the point when and where the entity ceases to be. As a result, a singular space-time path must exist provided an entity exists. However, it does not necessarily mean that a space-time path must have a longitudinal linear configuration. On the contrary, space-time paths can have either longitudinal or branching configurations (Figure 2.6).

Longitudinal configurations imply there is no possibility of having two or more directions over the past or future during a lifespan of an entity. They also imply that the exact spatial location of an entity is known at all times. Most cases of short-term changes making up the history of day-to-day occurrences of an entity portray longitudinal space-time paths. Examples are transport maintenance of state roadways, public works of utility companies, and cadastral measurements. On the other hand, branching configurations are encountered in medium- and long-term changes in the lifespan of an entity. Examples are the effects of pollution given various climatic and economic scenarios, accident and disease patterns, and land use and demographic trends.

Each space-time path is in fact the spatio-temporal signature of any entity in a knowledge domain. And as such, it has the ability to reveal the connections and interrelations among different entities. The identity, location and direction of a space-time path are fundamental structures to connect and interrelate the spatio-temporal data categories (state, event, episode and evidence). The following sections describe how this can be achieved.

2.10.1 State as an element of the space-time path

A state can be depicted as a specific location in the space axis, illustrated here using two dimensions for clarity. In the lifespan of an entity, changes occur in its space-time path describing the evolution of its inner existence, or a mutation of its location in space. Historical evolution regards change as the adaptation of an entity to

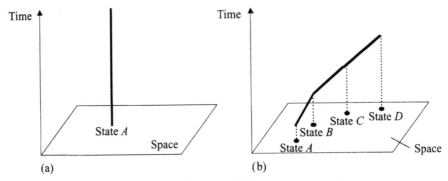

Figure 2.7   Examples of states as elements of the space-time path

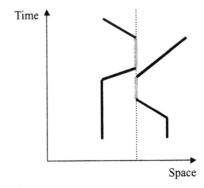

Figure 2.8   Co-location in space

its environment by the process of differentiation and increasing structural complexity (Figure 2.7(a)). Mutation relates change to the theory of revolution, emphasising the importance of conflict or struggle as the principal mechanism of change. It characterises any alteration in the direction of the space-time path, hence the change of its location in space (Figure 2.7(b)).

States tend to represent the short-term changes which can occur during the lifespan of entities. Such changes can occur due to man-made alteration in the position of entities on the ground. In contrast, states can also be related to the long-term and medium-term changes such as environmental changes. Changes can also occur due to natural causes, and the most common example is displacement of watercourse for rivers and streams. In any of these cases the spatio-temporal path gives the co-location of changes in space and time, as illustrated in Figure 2.8.

2.10.2   Event as an element of the space-time path

Those familiar with event-oriented representation and update-oriented representation should not mistake the concept of event in Time Geography. Both representations

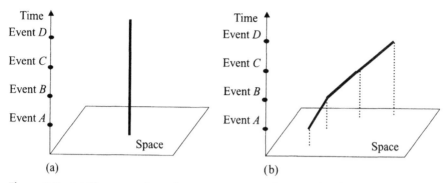

Figure 2.9 (a) The space-time shows a sequence of events that have occurred over the lifespan of an entity without causing any change in the spatial location of this entity. (b) The space-time path shows the ordering of the events and their respective associated changes over the spatial location of an entity

have been discussed as pragmatic solutions for representing spatio-temporal data in GIS (Langran, 1993). In an event-oriented representation, events are described by the moments of change. Only the events accountable for some sort of change over a state of an entity are represented. The occurrence of an event always causes the creation of a new version from a previous state of an entity. Conversely, in an update-oriented representation, events are related to the occurrence of updates of the data stored in a GIS. The lifespan of an entity is represented by a sequence of events that represent the occurrence of all kinds of updates, even those updates that are not accountable for any change in the state of this entity. These updates usually constitute a resurvey of an entire region, regardless of where or what change has occurred. Photogrammetric and remote sensing surveys are examples of collection-driven updates that are scheduled to occur in a given time interval.

A different perspective in representing events is found in the space-time path described in Time Geography. Events are not necessarily related to the creation of versions (states) or update activities in GIS. Time Geography emphasises the need for understanding geographic processes in which the mechanisms of change as well as the patterns of change have to be analysed through time. The sequence of events through time is viewed as the spatio-temporal manifestation of certain processes. The space-time path provides the connection, ordering and synchronisation of events, and their association with the respective changes (states) if they have occurred over a lifespan of an entity (Figure 2.9).

In placing events in the time axis of a space-time path, the meaning of an existence and mutation is given to a lifespan of an entity. Each space-time path captures the spatial and temporal sequence and coexistence of events. An ensemble of space-time paths belonging to several different entities can also represent the interaction among these entities. Such an interaction is given by the co-location of events in time (Figure 2.10). GIS can be used to keep the record of 'co-location of events' of space-time paths of a number of entities. The outcome is a web formed by the interrelations of individual trajectories of several space-time paths. Such

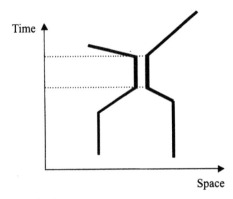

Figure 2.10   Co-location in time

a web can uncover processes that are responsible for the 'temporal connectedness' of common existence in time. Hägerstrand (1975) has previously named these processes as 'collateral processes' – processes which do not unfold independently but are observed from their common existence in time.

2.10.3   Episode as an element of the space-time path

An episode is defined as the length of time during which change occurs, a state exists or an event lasts. The length of the space-time path and its angularity with the time axis are important in classifying changes according to their respective duration: the larger the angle, the shorter the duration. Three main types of change can be characterised in a space-time path (Parkes and Thrift, 1980):

- Long-term changes modifying the environment
- Medium-term changes transforming cultures
- Short-term changes making up the history of day-to-day incidents

2.10.4   Evidence as an element of the space-time path

An entity is represented as a point object in Time Geography. Evidence is the data about entities, events and states that have occurred. Each place on the space-time path can only exist if there is a state associated with that place and evidence to confirm an event occurred (Figure 2.11). The space and time axes will determine the scale in which a space-time path occurs. In practice this is determined by the spatio-temporal data collected for a GIS.

2.11   UNCOVERING SPACE-TIME PATHS

The possibilities of defining the events, states and episodes that belong to a space-time path are immense for any knowledge domain in geographic information

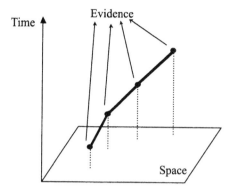

Figure 2.11 Evidence as an element of the space-time path

sciences. Consequently, a space-time path can be constructed according to an ensemble of *constraints* which define the circumstances what, where and when over space and time. These constraints operate on entities, events and states depending on a set of circumstances linked to the individual entity and its environment. This involves spatio-temporal data modelling of the behaviour of space-time paths.

Three types of constraint are described in Time Geography (Hägerstrand, 1975) and can be used in designing the spatio-temporal data model:

- *Capability constraints* limit the trajectories of the space-time paths. Space has a limited capacity to accommodate events because entities cannot occupy the same space at the same time, and every entity has a geometric boundary in space. Therefore every space has a packing capacity defined by the types of entities to be packed into it. Some constraints have a predominant time factor, at rather strictly regular intervals. Others can have a dominant space factor, forming bounded regions or volumes.

- *Coupling constraints* define where, when and for how long the events and states have to join a space-time path of an entity. Coupling constraints can reveal the pattern of space-time paths by exhibiting prism, area or volume configurations known as the potential path areas (PPAs).

- *Authority constraints* impose restricted access to space-time paths. The main purpose of defining authority constraints is to organise access to the data as well as to define domains of authority. In Time Geography the domain of authority is shown as a cylinder (Figure 2.5).

2.12 CONCLUSIONS

This chapter has focused on the main concepts of time geography. Many references in the literature introduce the Time Geography approach and its possible applications. The reader interested in an expanded coverage of these topics can find a good start with Hall (1966), Karlqvist, Lunqvist and Snickars (1975), Lenntorp (1976),

Pred (1977), Carlstein, Parkes and Thrift (1978), Parkes and Thrift (1980) and Golledge and Stimson (1997).

The main concepts of Time Geography discussed in this chapter play an important role in the design of spatio-temporal data models in GIS. Time Geography is not just another theoretical framework but it discloses how the integration between the absolute and relative views of space and time can be devised in GIS. Besides, the whole rationale of the Time Geography approach is the inseparability of space and time. In other words, states, events, episodes and evidence are all interrelated and connected through a space-time path of an entity. The space-time paths provide the image schemata for modelling spatio-temporal data of a knowledge domain in GIS. Very little information is available in the published literature on using time geography with GIS. This chapter has described two efforts made in this area, Miller (1991) and Makin (1992).

The semantics of states, events, episodes and evidence used in this chapter, in order to describe a space-time path, were not related to any particular level of abstraction, be it geographic (nation, region, centre), temporal (year, month, day) or demographic (population, group, individual). The purpose was to provide suitable semantics for modelling spatio-temporal data at multi-scales in space and time; readers may then adapt them to their specific knowledge domains. A practical example will be given in Chapter 5 to show how these semantics can be applied and implemented into a spatio-temporal data model based on knowledge domain areas such as historical geography.

The usefulness of Time Geography lies in providing space-time semantics to object-oriented analysis and design of spatio-temporal data models in GIS. Chapter 3 considers the concepts behind the object-oriented approach.

CHAPTER THREE

Object-oriented analysis and design

Object-oriented methods cover methods for design and methods for analysis. Sometimes there is an overlap, and it is really an idealization to say that they are completely separate activities.
I. Graham, *Object-Oriented Methods*

This chapter provides a historical background to object-oriented data management, illustrating the diverse efforts involved in object-oriented methods, temporal databases and version management approaches. It helps to explain the main concepts in the object-oriented paradigm that are essential for developing a spatio-temporal data model. A historical background on object orientation summarises the chronological developments from object-oriented programming languages to object-oriented design methods, and finally to object-oriented analysis methods. It can be difficult to choose which object-oriented method to apply in a spatio-temporal data model. For integrating the time geography framework within our spatio-temporal data model, the object-oriented analysis and design method proposed by Booch (1986, 1991, 1994) is presented as the best in terms of notation, completeness and technique.

The temporal database research is reviewed on the basis of concepts and techniques developed to establish appropriate temporal data management support for a spatio-temporal data model. Version management approaches are then described, emphasising approaches for ordering and updating versions within a model. The version management approach should be chosen so it can be effectively integrated with the spatio-temporal data model. In our spatio-temporal data model, *versions* are deemed to be distinct from snapshot series because they represent *states* that belong to the space-time path of an entity.

3.1 HISTORY OF THE OBJECT-ORIENTED PARADIGM

The history of object orientation starts in the early 1960s with the efforts of Dahl, Myrhaug and Nygaard in creating and implementing new concepts for programming discrete simulation applications. By 1965 the object-oriented programming language Simula (Dahl and Nygaard, 1966) had been developed on the basis of the ALGOL-60 language, which was specifically oriented towards discrete event simulation. Later, in 1967, the same Norwegian team developed the programming language Simula-67 (Dahl, Myrhaug and Nygaard, 1968), once again an extension of ALGOL-60. It is with Simula-67 that the basic concepts which characterise existing object-oriented programming languages were first introduced. In particular, Simula-67 introduced the notion of an object class defined by its type and the algorithms necessary to its manipulation. It also introduced the inheritance mechanisms through which an object class could inherit the data and the algorithms from other object classes.

However, it was only after the mid-1970s that the concepts introduced by Simula-67 were widely recognised. The programming language Smalltalk, a result of the work accomplished by Kay, Goldberg, Ingals and others at the Xerox Research Center at Palo Alto (PARC), has become established as the purest representation of object-oriented concepts. In Smalltalk everything is perceived as an object, and objects communicate with each other by passing messages. Having its origins in Simula and the doctoral research work of Alan Kay, Smalltalk has evolved by integrating the notion of classes and inheritance from Simula as well as the functional abstractions flavour of LISP.[1]

There have been five releases of Smalltalk running from Smalltalk-72, launched in 1972, to Smalltalk-80, launched in 1980; the other three releases were launched in 1974, 1976 and 1980. Smalltalk-V and Smalltalk-AT have also been created as dialects from the former Smalltalk developments (Krasner, 1981). Generally, Smalltalk is a complete programming environment, having features such as editors, a class hierarchy, browsers and many of the features of a fourth-generation language (Graham, 1994). Booch puts it like this:

> Next to Simula, Smalltalk is perhaps the most important object-oriented programming language, because its concepts have influenced not only the design of almost every subsequent object-oriented programming language, but also the look and feel of graphic user interfaces such as the Macintosh user interface and Motif.
>
> (Booch, 1994, p. 474)

Several object-oriented programming languages have been developed, most of them having their conceptual foundations based on Smalltalk. These attempts have tried to overcome the main inefficiency problems of Smalltalk (e.g. the lack of support for persistent objects and unfeasibility of having a distributed multi-user environment) but with the pitfall of compromising the purity and consistency of Smalltalk's features. Over 100 object-oriented programming languages have been developed in the past decade. However, as Stroustrup points out: 'One language is not necessarily better than another because it has a feature the other does not – there are many examples to the contrary. The important issue is not how many features a language

---

[1] LISP stands for list processing; it was originally developed by John McCarthy in 1958 and more recently it has been used in artificial intelligence work.

has, but that the features it does have are sufficient to support the desired programming styles in the desired application areas' (1988, p. 11). Object-oriented programming languages are still being developed and it is expected that new languages will emerge, acquiring new features rapidly.

From the mid-1970s onwards the research and development in artificial intelligence (AI) programming environments have also influenced the object-oriented paradigm. LISP is one of the main programming languages used in AI systems, and several object-oriented extensions of LISP have been created. LOOPS, Common LOOPS, FLAVOURS, KEE, ART and New FLAVOURS are some examples in which a semantically ample form of inheritance is proposed that differs from the one encountered in most object-oriented programming languages such as Smalltalk. Here values, in particular default values, can be inherited as well as attribute names. Graham assures his readers that, from his point of view, 'AI people have got it right and that this kind of inheritance will gradually penetrate the world of object-oriented programming' (1994, p. 78).

With the maturing of the concepts in object-oriented programming languages and their practical use in different application contexts, research interests have diversified, focusing on object-oriented design methods:

> Object-oriented design is a method of design encompassing the process of object-oriented decomposition and a notation for depicting both logical (class and object structure) and physical (module and process architecture) as well as static and dynamic models of the system under design. (Booch, 1994, p. 39)

Significant debate has occurred in this research area, mainly concerning whether an object-oriented design method can be intrinsically independent of any programming language, or whether current design methods are clearly attached to specific object-oriented programming languages. Most object-oriented design methods reveal the influence of Booch's pioneering work (Booch, 1986). In his original proposal, Booch suggested a design method based upon some features of the ADA programming language, using an object-oriented style. GOOD[2] and HOOD[3] are examples of ADA-derived methods that enforce the top-down hierarchical decomposition approach among objects but without supporting inheritance and polymorphism.

Also influenced by Booch's work, OOSD[4] provides a hybrid, low-level notation for logical design of object-oriented methods in general. Although designed to be language independent, OOSD has not been extended to a consistent object-oriented notation due to its inability to deal with complex data structures and large numbers of methods. OODLE[5] is another example of a language-independent notation which advocates four interrelated diagrams in order to support the Shlaer–Mellor approach to object-oriented design. Booch's revised design method (Booch, 1991,

---

[2] General Object-Oriented Design method developed at NASA.
[3] Hierarchical Object-Oriented Design method developed at the European Space Agency.
[4] Object-Oriented Structure Design introduced by Wasserman, Pircher and Muller (1990).
[5] Object-Oriented Design LanguagE is a design-specific component of the Shlaer–Mellor method (Shlaer and Mellor, 1988).

1994) gives probably the most incisive and comprehensive prospect of an object-oriented design method. His method improves the concepts of object orientation and their respective notations as a whole, overlapping with the concepts of object-oriented analysis.

Other research innovations have emerged from the synergy between object-oriented programming and database management systems. This has generated a potential mechanism for representing, storing, organising, sharing and recovering objects that include multiple complex data types and associated methods and functions. Object-oriented database systems (OODBS) have developed capabilities such as persistence, long transactions and versioning, unlike most traditional relational database management systems. Through combining database functionalities with object-oriented programming, OODBS has become an expressive device for multimedia applications, client–server systems as well as GIS, CAD, engineering and manufacturing systems.

Object-oriented databases have emerged as commercial products. ONTOS,[6] O2,[7] GemStone,[8] ObjectStore[9] and ORION[10] are some examples of object-oriented databases, although their capabilities can differ widely. These object-oriented databases have in common basic characteristics such as methods associated with objects, inheritance of attributes and procedures from supertypes (superclasses), and the ability to define the type (class) of objects, their attribute types and relationships. However, they differ substantially in their query languages. The enormous differences probably result from the fact that OODBS have been elaborated based on programming languages for their data models. Sometimes declarative query languages are only introduced after the initial implementation. The lack of a standard or a formal background for object-oriented query languages has caused differences in query language syntax, completeness, SQL compatibility and treatment of encapsulation (Cattell, 1991). Graham puts it like this:

> The most recent geographic information systems, such as Smallworld, have opted for an object-oriented approach to storing mapping data. The authors of Smallworld chose to create their own persistent version of Smalltalk and object-oriented database because no commercial OODBS existed at the time they started. Vendors starting now have a much better choice.   (Graham, 1994, p. 117)

Object-oriented databases also offer the possibility of storing and manipulating all data pertaining to a GIS application in the same manner. By contrast, in relational databases, spatial data cannot be so readily stored and their integration with other systems is cumbersome. Chance, Newell and Theriault (1990) advocate the benefits of object-oriented concepts in developing a seamless environment. In the case of Smallworld GIS, object-oriented database capabilities have been implemented by

---

[6] ONTOS is a product of Ontologic, Billerica MA, which enhances C++ with persistent objects.
[7] A commercial product of GIP Altair, Le Chesnay, France. It reveals strong Prolog influences.
[8] A product of Servio Corporation, Alameda CA and Beaverton OR. It has been built onto an extension of Smalltalk-80 known as OPAL.
[9] A product of Object Design, Burlington MA, based on the C++ programming language.
[10] A commercial product of Itasca Systems, Minneapolis MN, which extends LISP with object-oriented capabilities.

front-ending a version-managed tabular data store with an object-oriented language named Magik. In this environment, system programming, applications development, system integration and customisation are all written using the same object-oriented programming language, i.e. Magik. 'Object-orientation does not just mean that there is a database with objects in it, but that the system is organised around the concept of objects which have behaviour (methods)' (Chance, Newell and Theriault, 1990, p. 181).

The question arises as to whether existing relational database products will be superseded or whether they will evolve into some sort of extensions to include object-oriented concepts such as methods, object identity, complex objects and object versions. This has raised a number of issues concerning the relative efficiency of declarative relational query languages and the unfeasibility of storing processing logic at the table level. POSTGRES[11] and Starburst[12] are representative examples of the most advanced implementations of extended relational databases aiming at the main object-oriented features. According to Cattell:

> There is no single 'extended relational approach', in the sense of DBMSs built on the relational model with a common query language and model; indeed, there may be less standardization and consistency than in some other ODMS [object data management systems]. However, the extended relational approach is popular because . . . [it] can benefit from much of what has been learned about relational systems, and, more important, it may be possible to migrate users from existing relational database products to . . . [extended relational databases]. (Cattell, 1991, p. 83)

Following the proliferation of research on object-oriented programming and database management systems, object-oriented analysis methods have been gradually developed as an approach to improving our understanding of the concepts, activities, rules and assertions of the object orientation paradigm. 'Object-oriented analysis is a method of analysis that examines requirements from the perspective of the classes and objects found in the vocabulary of the problem domain' (Booch, 1994, p. 39). Within the object orientation paradigm, methods developed for object-oriented design are frequently applicable to object-oriented analysis, and vice versa. Computer-aided software engineering (CASE) has become increasingly important as a graphical tool for supporting object-oriented analysis and design methods. CASE tools have been variously regarded with enthusiasm or with disbelief that there is any advantage to be gained through their use.

An increasing number of software products for CASE tools are under development based on the composition of graphical symbols and notations depicting the semantics and features from the object-oriented analysis and design methods. The most important benefits of using CASE tools are their ability to generate code automatically and enhance productivity. However, CASE tools can restrict innovative kinds of application, where the rules and methods provided by CASE tools are

---

[11] POSTGRES, from the University of California at Berkeley, is an extension of INGRES with objects, multiple inheritance, versions, historical data and a powerful extended relational query language (INGRES QUEL).

[12] Developed at the IBM Almaden Research Center, it extends the relational algebra.

inappropriate or even non-existent. The main examples of CASE tool systems available in several platforms and operating systems are the ROSE tool supporting Booch's method; Object Maker with support for a vast range of conventional and object-oriented methods, including Booch, Coad–Yourdon, Shlaer–Mellor, Rumbaugh and HOOD; and OOATool supporting the Coad–Yourdon method.

The current stage in the history of the object-oriented paradigm is characterised by an awareness of the issues related to the necessity of developing open systems and standards. The Object Management Group (OMG), formed by several industry representatives, has undertaken the task of reaching an industry-wide consensus for a reference architecture and data model for object-oriented database management systems (OODBMS). As an organisation in charge of creating and promoting a standard for OODBMS, OMG has proposed the object database standard ODMG-93 1.0 which specifies an ODM (object data model), ODL (object definition language), OQL (object query language) as well as C++ and Smalltalk language bindings for OODBMS. Conforming to the ODMG-93 standard, an object-oriented database might supply tools for implementing ODM, ODL and OQL features. The ANSI SPARC OODB Task Group has also been working on a reference object-oriented data model (NIST, 1991; ODMG, 1994; Kim, 1991). Although the ANSI Standards Committee has not yet recognised the ODMG-93 as an official standard, they have begun to define some references for object information systems (ANSI X3H7) and managed objects (ANSI X3T5.4). Finally, the development of open distributed processing environments (CORBA,[13] OLE[14] and DCE[15]) has introduced a revolutionary approach that is totally centred in a client–server architecture for databases. These architectures consist of an integrated set of technologies that make it easy to create, use and maintain applications in a distributed environment. Most of the ODBMS vendors claim they can provide an efficient client–server architecture for their databases.

3.2 CHOOSING AN OBJECT-ORIENTED METHOD

'Is there a "best" [analysis and] design method?' Booch (1994, p. 23). Booch's conclusion was: 'No, there is no absolute answer to this question.' Nevertheless, the decision to apply an object-oriented analysis and design method to developing the spatio-temporal data model, was motivated by the simplicity as well as the complexity presented in this methodology. Due to its simplicity, the underlying ideas of the time geography framework can be described more easily using the primary concepts developed in object orientation. Also, the complexity of implementing the time geographic elements of such a model is an interesting challenge from both the conceptual viewpoint and the implementation viewpoint.

[13] CORBA (Common Object Request Broker) is from OMG (Object Management Group) and consists of an object model in which all communication is performed by the ORB middleware.
[14] Microsoft introduced OLE (Object Linking and Embedding) for integrating multiple applications and multimedia data types within a document framework.
[15] DCE (Distributed Computing Environment) from the Open Software Foundation is probably the most significant open systems standard for heterogeneous client–server interoperability.

Some main criteria have been established a priori for choosing a suitable object-oriented method without any particular solution in mind. These criteria are preconditions that should be fulfilled by the favoured object-oriented method to increase the quality, flexibility and clarity of the system. One of the criteria that inevitably arises as a management issue is the need to support an extremely rich notation to describe the main concepts of object orientation which will be employed to model the spatio-temporal representation of Time Geography. However, this notation must not be complicated or difficult to employ, to allow an easy learning and understanding of the layers, phases or activities pertaining to the object-oriented method. A good analogy is the entity–relationship (ER) diagram (Chen, 1976) which is the most common approach to relational data modelling due to the clearness and comprehensibility of its graphical notation. This notation can immediately be translated into a database implementation.

Another complementary criterion is the need to represent the dynamics of change within the object-oriented methodology. There is a need for handling rules (topological rules), constraints (capacity, coupling and authority constraints), and methods (trigger methods – update, delete, create) at the levels of both object and class. The graphical notation might show a scenario with a message being passed from one object to another as well as ensuring that the message is in fact part of the protocol of an object. Finally, an implementation criterion has also been taken into account, namely language independence. The object-oriented method has to be language independent, which means it cannot be restricted to a specific object-oriented language such as C++ or Smalltalk. This is paramount for assuring an overall integration in the system between the stages of analysis, design and programming.

The object-oriented method proposed by Booch (1994) is a major contribution to unifying ideas by incorporating the best from each of the existing object-oriented methods, including the work of Jacobson, Rumbaugh, Coad and Yourdon, Constantine, Shlaer and Mellor, Firesmith and others. A unified notation has been achieved in which the 'cosmetic differences' between Booch's notation and those of other object-oriented methods have been reduced, particularly the notation used by Rumbaugh in the OMT method. Booch asserts that 'Rumbaugh's work is particularly interesting, for as he points out, our methods are more similar than they are different' (1994, p. vi).

Four distinct models are defined to produce the analysis and design within an object-oriented development, namely the logical model, the physical model, the static model and the dynamic model. These models are then grouped into two dimensions: the logical/physical view and the static/dynamic view. For each dimension, Booch has defined a number of diagrams which denote a view of the models of the system. The class diagram, the object diagram, the module diagram and the process diagram are used to capture the semantics within the logical and physical models.

In the case of the static and dynamic models, two additional diagrams are proposed: state transition diagrams and interaction diagrams. Each class may have an associated state transition diagram which indicates the event-ordered behaviour of the objects. Similarly, in conjunction with an object diagram, an interaction diagram can be provided to show the time ordering of messages. Booch presents a fairly rich notation as well, but one that is easier to understand, hence easier to apply. For example, the notation used in the interaction diagram is actually a generalisation of event diagrams of the dynamic model of OMT (Rumbaugh *et al.*, 1991) combined with the interaction diagrams of Jacobson's method (Jacobson *et al.*, 1992).

Booch's method distinguishes two processes in object-oriented development, the micro process and the macro process. The micro process is more closely related to the spiral model of Boehm (1986) and serves as the framework for an iterative and incremental approach to development. The macro process is more closely related to the traditional waterfall life cycle and serves as the controlling framework for the micro process (Booch, 1994). This provides a flexible and legitimate object-oriented model of an application in which analysis and design techniques have been integrated for each process, model and view of the object-oriented development.

Booch argues that the processes within an object-oriented analysis and design method cannot be described in a 'cookbook'. His proposal for an object-oriented development embodies purpose, products and activities which are considered as incremental and interactive phases rather than steps. By avoiding a cookbook presentation, Booch emphasises processes, models and views within his object-oriented method. This provides a flexible and legitimate object-oriented model of an application in which analysis and design techniques have been integrated for each process, model and view of the object-oriented development.

Booch introduces two main concepts about objects in his object-oriented method. First there is the client–server concept between objects. A client object is an object that uses the operations of another object, either by operating upon it or by referencing its state. Conversely, the server object is the object which provides the operation. Second, the existence of a state within an object means the order in which operations are invoked is important. This raises the concept of active and passive objects. Active objects can manifest some state change without being operated upon by another object; they hold their own thread of control. Whereas passive objects can only undergo a state change when explicitly acted upon. In this manner, the active objects in the system serve as roots of control. If the problem domain involves multiple threads of control, we will usually have multiple active objects (Booch, 1994).

Four basic operations can act upon an object (Booch, 1994):

- *Modifier*: an operation that changes the state of an object.
- *Selector*: an operation that captures the state of an object, without changing the state.
- *Iterator*: an operation that allows access to some properties of an object's state in a well-defined order.
- *Destructor*: an operation that frees the state of an object and/or destroys the object itself.

The definition employed by Booch for behaviour also discerns that the state of an object affects its behaviour. In other words, objects do not have a static state, but each state of an object represents the cumulative results of its behaviour. At any

Table 3.1   Objects and some properties to represent them.

Object	Properties to represent that object
Spatial references	Coordinates, enclosing rectangles, place names or codes
Spatial relationships	Connectivity, orientation, adjacency, containment
Spatial factors	Scale, resolution, units of space, map projections
Spatial dimensions	Point, line, area, polygon, grid cell
Spatial measurements	Length, perimeter, surface area, volume, orientation
Thematic information	Qualitative and quantitative attribute values, descriptive information
Temporal references	Time stamps, valid time, transaction time

point in time, the state of an object involves all properties of this object (usually static) as well as the current values of these properties (usually dynamic). An object can have any sort of properties representing its state, as illustrated in Table 3.1.

As the understanding of a knowledge domain varies from one individual to another, objects can be referred to a common class or the same object can belong to different classes at the same time. Deciding upon the best classification for a given knowledge domain is a fundamental aspect of object-oriented analysis and design. Burrough asserts that classification 'is essential for human understanding – without classification or generalization our brains become swamped by detail' (1986, p. 137). Booch devotes a chapter to the subject of classification in which three approaches are described: classical categorisation, conceptual clustering and prototype theory.

As a main criterion in its classification process, *classical categorisation* uses the association of common behaviour or properties among objects to categorise the object classes. Two main assumptions are taken in the classical approach:

- Classes are like containers, with objects either inside or outside.
- Objects in the same class must have the same properties.

MacEachren (1995) points out the use of classical categorisation in cartography. Choropleth maps, soil maps and climate maps are examples of classical categorisation. In these examples, objects (map units) belong to a particular category and are depicted with identical symbolisation. In contrast, *conceptual clustering* attempts to group the conceptual descriptions of classes to which objects may belong. And here the main assumptions are as follows:

- Classes are defined by a membership criterion (or criteria).
- Objects in the same class do not necessarily have the same properties.

Bayesian classification is an example of employing statistical theory to obtain membership classifications of objects to multiple classes (Glymour *et al.*, 1997). For applications that have neither clear concepts nor clearly bounded properties and behaviour, *prototype theory* is considered as an alternative classification approach. In this case, objects are grouped according to their degree of relationship to concrete prototypes. Prototypes for a class are simply exemplars of most typical objects that can be found in that class. The classification is determined by similarity to a prototype.

Object-oriented analysis and design methods are based on classification structuring that is highly dynamic and knowledge domain dependent. There is no single correct approach to classify a phenomenon in a knowledge domain. MacEachren (1995) provides an interesting discussion on these classification approaches from a cartographic perspective. However, it is important to observe that the classical categorisation approach has been employed for designing our spatio-temporal data model.

3.3   THE MAIN MODELLING CONSTRUCTS

The main object-oriented modelling constructs required by the spatio-temporal data model are described in this section. They are knowledge domain, scenarios, objects, inheritance, classes, time dimension, methods and data model changes.

Knowledge domain

Defining the boundaries of a knowledge domain relies on (a) understanding the concepts that describe this knowledge domain and (b) identifying the concepts that are relevant to the scope of the spatio-temporal data model. The structure of a spatio-temporal data model depends largely on the view taken with respect to this domain. Thus, the structural compatibility between concepts and modelling constructs (e.g. objects, classes and scenarios) depends a great deal on the assumed view of the knowledge domain. Concepts provide an efficient means for understanding any knowledge domain. However, they are not fixed to any particular modelling construct. We construct our reality of interest by using concepts. We communicate to this reality by using modelling constructs.

Scenarios

Scenarios are integrated subsystems within a data model. They are task-driven conceptualisations of the knowledge domain. Depending on the specific task to be solved, each scenario isolates the relevant aspects of particular objects, classes, properties and/or relations.

Objects

A class is a set of objects that share common properties. A single object is simply an instance of a class. A unique identifier (OID) is always given to each object, independently of the value of its properties. Creation of a new object, creation of a new object from an existing object, and change in state of an existing object characterise the dynamic nature of objects within an object-oriented model. At any point in time, the state of an object involves all properties of this object (usually static) as well as the current values of these properties (usually dynamic). As a result, a complex structure of properties can be found in a class. Ahmed and Navathe (1991) have proposed a taxonomy to identify every property of an object as being one of the following attributes:

- Invariant attributes which cannot be modified or deleted at any time.
- Version significant attributes which can be updated only in a structured manner.
- Non-version significant attributes which can always be updated without creating a new version.

An update in a version significant attribute creates versions of an object bearing this change. Two types of update are possible in a data model, atomic and non-atomic. A non-atomic update modifies an object's attributes several at a time. Conversely, an atomic update modifies an object's attributes one at a time. Atomic updates are rarely used in GIS. This taxonomy is based on the four operations defined for an object (modifier, selector, iterator and destructor). The primary issues in object-oriented data analysis and design are to assist in choosing what should be versioned and to provide the mechanism for controlling unnecessary proliferation of versions.

Inheritance

Also called generalisation, subtyping or subclassing, inheritance represents a generalisation hierarchy in which the 'is a' relationship is among the classes. It emphasises the top-down approach with the most general superclass (parent) at the top and the most specific subclass (child) at the bottom. The subclasses inherit the state (properties) and behaviour (methods) from their superclasses, adding to them their own state and behaviour.

Classes

The iterative and exploratory nature of an object-oriented analysis and design method provides three system-defined classes of object. They are the basic modelling semantics for supporting version management in the object-oriented model. Therefore, each class in the model must pertain to one of the following types:

- *Generic class* identifies the class which represents the essence of a set of objects and contains a high level of abstraction in its functionality.
- *Versioned class* is a subclass of the generic class; it contains the versions of an object. Whenever an object is created or modified, versions of this object should be available to allow the use of multiple states of the same object. Three main issues can be related to versioned classes in object-oriented analysis and design: when to create a new version, how to represent the version, and which versions in a database represent a consistent configuration of objects.
- *Unversioned class* is a static class where its objects never change.

Time dimension

Snodgrass (1992) argues that any implementation of any data model with a temporal dimension will have to hold some discrete encoding of time. The time unit is called a chronon; a chronon is the smallest duration of time that can be represented in a data model. In object-oriented analysis and design it is possible to identify five fundamental types of discrete encoding of time:

- Nominal: today, 13 May
- Binary: earlier than, later than
- Ordinal: time ago, long ago
- Interval: event *A* is 1 year
- Ratio: event *A* starts at 8 and ends at 6

Methods

A method is a general program associated with a class and, when invoked by a message passing to the class, it is executed on all instances of that class. There is practically unlimited application of methods. The capability of objects from different classes to respond to the same message is called polymorphism. A protocol is the entire set of methods that may be performed upon an object.

Data model changes

Object-oriented data models are not static. Relatively complex changes can occur in the lifespan of most data models. Banerjee *et al.* (1987) propose the following taxonomy for evolution of a schema:

1   Changes to the components of a class
  (a)   Changes to properties
    - Add a new property
    - Drop a property
    - Change the name of a property
    - Change the type of a property
    - Inherit a different property definition
  (b)   Changes to methods
    - Add a new method
    - Drop a method
    - Change the name of a method
    - Change the implementation of a method
    - Inherit a different method definition
2   Changes to relationships of classes
    - Add a new class relationship
    - Remove a class relationship
    - Change the class relationship

3   Changes to classes themselves
- Add a new class
- Drop an existing class
- Change the name of a class

Data model changes describe the evolution of object-oriented data models that involve updates to the schema. They are not related to the update procedures on the data available in the database. They can be considered as changes that need to be performed in order to rectify errors encountered in the schema of a data model. Otherwise, they can also be related to the changes which are necessary due to the development of our understanding of the knowledge domain.

Data model changes play an important role in a distributed GIS environment with a large number of users. A version management mechanism has to be provided so that a designer can test any data model change to the data model in their own version of the whole database. Some GIS products present a mechanism for handling data model changes, e.g. Smallworld GIS. The CASE tool[16] in Smallworld GIS has been specifically developed for handling data model changes. The version management implemented in the CASE tool has the ability to define and modify object classes: 'This uses version management techniques extensively to help functions like the ability to create alternative versions of the data model, apply these to alternative versions of a populated database for testing and development, and finally apply the changes to the master version of the database with a mechanism for propagating the schema changes down to other versions in the database' (Newell and Batty, 1993, p. 3.2.3).

3.4   TEMPORAL DATABASES

By the end of the 1980s, researchers had recognised the need for databases to support information that varied in time (Ackoff, 1981; Anderson, 1982; Ben-Zvi, 1982; Bonczek, Holsapple and Whinston, 1981; Clifford and Warren, 1983; Snodgrass, 1987). However, certain misconceptions had arisen concerning the terminology and definition of the features supported by temporal databases. The work of Snodgrass and Ahn (1985) was the first to present a new taxonomy of time that unified the concepts developed in the literature.

Transaction time has been proposed as a consensus term for previously defined physical time (Dadam, Lum and Werner, 1984; Lum et al., 1984), registration time (Ben-Zvi, 1982), data valid from/to (Mueller and Steinbauer, 1983) and start/end time (Reed, 1978). Correspondingly, valid time has been proposed for event time (Copeland and Maier, 1984), effective time (Ben-Zvi, 1982), logical time (Dadam, Lum and Werner, 1984; Lum et al., 1984), state (Clifford and Warren, 1983) and start/end time (Jones, Mason and Stamper, 1979; Jones and Mason, 1980).

The terms 'transaction time' and 'valid time' have also been proposed to differentiate the distinct types of database according to their ability to handle temporal

---

[16] CASE tool is an interactive, visually oriented tool for creating and managing the database schema using graphical notations.

information. Four types of database have been defined according to their ability to support these time concepts and the processing of temporal information (Snodgrass and Ahn, 1986):

- *Snapshot databases* provide a unique snapshot of a database state.
- *Rollback databases* provide a sequence of snapshot states of the database by offering support for transaction time (the time the data are stored in the database).
- *Historical databases* provide the knowledge about the past by supporting valid time (the time the data have changed in the real world).
- *Temporal databases* support both transaction time and valid time.

Using the taxonomy of time employed in the research on temporal database systems, object-oriented databases can be considered as rollback databases due to their ability to represent temporal information. Object-oriented databases support the history of the transaction results in snapshot states rather than the history of the real world as we perceive it. Most object-oriented databases store all past states, indexed by transaction time, of the snapshot database as it evolves. Object-oriented databases may in the near future advance into the realms of temporal databases.

POSTGRES embodies the first substantive proposal for implementing a rollback database using optical disks with support for transaction management and concurrency control mechanisms (Stonebraker, 1987). In providing these extended database functionalities, the possibility of implementing spatio-temporal constructs as extended database functionality for POSTGRES has also raised expectations in the GIS field. The GEO System (van Hoop and van Oosterom, 1992) is an example of a prototype design in which spatial constructs have been implemented by developing some extensions to the POSTGRES structure. Probably the best-known effort to implement a spatio-temporal data model using the POSTGRES database is the Sequoia 2000 Project, led by Michael Stonebraker of the University of California at Berkeley (Gardels, 1992). POSTGRES has been integrated with the GRASS[17] GIS to allow temporal manipulation of global change data.

The temporal database management system (DBMS) prototype developed by the IBM Heidelberg Scientific Centre, in the Advanced Information Management Project, was the first attempt to implement both valid time and transaction time with the support of temporal indexing (Dadam, Lum and Werner, 1984; Lum *et al.*, 1984). This attempt has been important in consolidating the concepts developed by temporal research in databases, and it has made people aware of the need for spatio-temporal indexing in the GIS field. For example, Langran (1988) pointed out spatio-temporal indexing as a temporal GIS design trade-off, in order to define how to control the volume of spatio-temporal data required in an application within a GIS. To achieve spatio-temporal indexing for a temporal GIS, Hazelton (Hazelton, Leahy and Williamson, 1990) discusses a natural extension of a quadtree structure; and most appropriate, he suggests, could be a multidimensional indexing structure.

---

[17] Geographic Resources Analysis Support System.

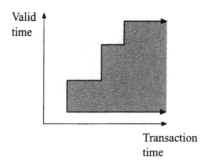

Figure 3.1   The bitemporal element

Snodgrass and Ahn (1985, 1986) have also demonstrated the existence of a true orthogonality between transaction time and valid time which allows each kind of time to be pursued independently: 'Time is a universal attribute in most information management applications and deserves special treatment as such' (Snodgrass and Ahn, 1986, p. 35). As a result, Snodgrass (1992) subsequently introduced the concept of a bitemporal element. This element is deemed to represent the valid time–transaction time space (Figure 3.1).

Several applications in GIS require information on the state of an object when it has been modified in the database to match its change in the real world. Worboys (1994) investigated the possibility of creating bitemporal elements for spatial chains (line objects) in object-oriented databases. The implementation used a discrete model for linear time. His findings show that the critical areas for research are the development of appropriate spatio-temporal data models, the indexing technique for this model, and related performance issues.

By looking for innovative directions in the temporal research domain, Snodgrass (1990) alludes to the outstanding potential of object-oriented databases which offer significant support for handling time, despite the lack of research work carried out in object-oriented databases about valid time. The important role of object orientation has also been identified in the proceedings of the first temporal GIS workshop (Barrera, Frank and Al-Taha, 1991), and the need for object versions to have an identity is deemed a fundamental temporal issue. The conclusions reached in the workshop were that the identity approach supported by object-oriented databases would be better suited for incorporating time into a GIS rather than the value-based approach of the relational database model. The challenge is to identify the capabilities provided in object orientation for the creation, modification and version maintenance of objects.

Kemp and Kowalczyk (1994) presented the main effort in incorporating a temporal capability within a GIS using the object-oriented paradigm. They employed object-oriented constructs as the tool for providing a flexible and adaptable temporal capability within a GIS. Johnson and Kemp (1995) have also provided a description of the implementation aspects of temporal capabilities within a GIS such as the use of functions (methods) to support spatial and temporal data types and operators.

Table 3.2 GIS version management: an overview.

Level of versioning	Approach	References	Query language	Pros	Cons
Raster cell	Linked lists of attributes for each cell	Langran (1992a)	N/A	Fairly simple approach	Static snapshots
Block	Block sharing in a version-managed B-tree	Newell, Theriault and Easterfield (1994)	Proprietary	Multiple versions of the same object	Designed primarily for long transactions
Relation	Table versioning	Clifford and Ariav (1986)	SQL	Simple and complies with relational algebra	High degree of redundancy
Tuple/object	Time stamps are associated to each tuple	Snodgrass and Ahn (1985)	Extensions to SQL	Efficient storage of versions	Difficulty with temporal joins
Tuple/object	Current tuples are stored in one table, historical tuples in another, linked by pointers	Lum et al. (1984)	Extensions to SQL	More efficient storage of versions	Historical queries are computationally intensive
Tuple/object	Time-change objects	Ramachandran (1992)	N/A	Flexible, object-oriented modelling potential	Non-standard database architecture requirements
Tuple/object	A bitemporal element is associated to each object	Worboys (1994)	N/A	Integration of transaction time and valid time	Non-standard database architecture requirements
Attribute	Time stamps are associated to each attribute	Clifford and Ariav (1986)	Extensions to SQL	Low level of redundancy	Non-standard database architecture requirements

3.5   VERSION MANAGEMENT APPROACHES

According to Loomis, 'Versioning is the tracking of the evolution of an object's state through time' (1992, p. 40). Deciding on a version management approach for a spatio-temporal data model involves finding a way to integrate the approach and the model. The version management approach requires being consistent with object-oriented concepts. Table 3.2 describes the main approaches developed for managing versions within databases and their application in a GIS context.

Currently, two main strategies are used to represent multiple versions of an object within an object-oriented database. The first strategy consists of version numbers or time stamps being associated with every attribute or relationship of an object. These attributes or relationships may then be chained, giving them a historical order. The result is a single object and identity (OID) with the versions actually associated with the attributes. This approach could have advantages only in models in which the objects have several attributes as well as changes occurring to only a few attributes.

The second strategy is to track versions at the level of objects rather than attributes. Chaining the old object and any older versions to the new version creates a new version of an object. As a result, a different identifier (OID) is associated with the new version, hence each version has its unique identifier within the version configuration. So far no performance studies have been undertaken on these alternatives. Nevertheless, both alternatives involve the design of operations for handling the dynamic behaviour of versions within the object-oriented data model. Ordering versions within a data model is an important aspect in object orientation. Despite the development of storage devices, such as optical disks, offering new capabilities for storing large amounts of data, a significant waste of storage space will still be minimised by ordering versions compactly. The proper choice of a method for ordering temporal data relies on producing the best performance for a given application.

3.6   CONCLUSIONS

Object orientation is an important related interdisciplinary research domain that can uphold the task of incorporating time in GIS. The fundamental concepts encountered in the object-oriented approach offer useful improvements in functionality, clarity of data modelling and the potential for simplifying future application developments in GIS: 'A common difficulty in . . . [GIS] application areas is the gulf between the richness of the knowledge structures in the application domains and the relative simplicity of the data model in which these structures can be expressed and manipulated. Object-oriented models have the facilities to express more readily the knowledge structure of the original application' (Worboys, Hearnshaw and Maguire, 1990, p. 370). Therefore, the object-oriented paradigm is viewed as an approach capable of handling the spatio-temporal semantics of Time Geography. The Time Geography framework will be embedded within our spatio-temporal model through the modelling capabilities provided by object orientation. Our spatio-temporal data model is described in the next chapter.

CHAPTER FOUR

The spatio-temporal data model

The term representation means an arrangement or organization of data defined within an explicit set of primitive elements [entities], attributes [properties], and relationships. Such an arrangement serves to preserve the information inherent in the data for subsequent use in problem-solving or analytical evaluation. A representation can be purely conceptual, but any representation must be expressed in formalized mathematical or programming terms for computer implementation. A formalized representation intended for computer implementation is known as a *data model*.

D. Peuquet

Developing our spatio-temporal data model (STDM) based on Time Geography and object orientation has certain implications that must be taken into consideration independently of the knowledge domain we are hoping to represent:

- The spatio-temporal data model emphasises the interaction between events and states rather than a linear connection among them over a space-time path.

- The space-time path offers an appropriate semantic abstraction for handling only valid time in a spatio-temporal data model.

- The space-time path requires a different version management approach for incorporating change into a database. Most of the approaches extend the relational model by creating new versions of tables, tuples or attributes to reflect changes occurring over time. A new version management approach is necessary to deal with changes that are represented by the synchronisation of events and states.

- Object-oriented analysis and design is required for the development and implementation of the spatio-temporal data model within a GIS.

The strengths of the STDM and its implementation lie in the well-defined concepts and representations developed in Time Geography and object-oriented approaches. The spatio-temporal data model offers much practical guidance, and implies the feasibility of applying object-oriented methods to the problem of handling space and time in GIS. This chapter gives a detailed description of the STDM used in the rest of this book.

4.1   DEFINING THE REASONING TASK

Spatio-temporal data modelling is about organising abstract concepts of a knowledge domain into formal descriptions. A spatio-temporal data model should provide constructs for representing a knowledge domain in terms of entity/event abstractions, relationship abstractions, behaviour specifications and interaction descriptions. The STDM is primarily constructed from the viewpoint of the fused space and time dimensions rather than the dimensions themselves and their orthogonality. The space-time path is treated as the core of the manifestation of space and time within this model. Therefore, each entity (real-world phenomenon, feature or object) from a knowledge domain has its own space-time path that represents its lifespan. However, the possibility of describing more than one space-time path for a singular entity is conceivable and can demonstrate the complexity that one can achieve in the STDM configuration. Shoham and Goyal (1988) distinguish four different reasoning tasks for developing space-time paths within an STDM:

- *Prediction* Given a description of the world over some period of time, and the set of rules governing change, predict the world at some future time.

- *Explanation* Given a description of the world over some period of time and the rules governing the change, produce a description of the world at some earlier time.

- *Learning new rules* Given a description of the world at different times, produce the rules governing change which account for the observed regularities in the world.

- *Planning* Given a description of some desired state of the world over some period of time and given the rules governing change, produce a sequence of actions that would result in a world fitting that description.

The first step towards designing space-time paths within an STDM is to uncover which task is deemed to be achieved. These tasks represent the overall goals associated with space-time paths of the STDM, rather than a delineation of the appropriate elements of a particular space-time path. The key issue here is to understand the spatial, temporal and semantic aspects of the data (or the knowledge domain) in relation to space-time paths. How data are expected to be explored in a search for spatial, temporal and spatio-temporal patterns is vital when delineating a conceptual configuration for designing space-time paths within the STDM; this chapter uses the second task – explanation.

Table 4.1 Main abstractions defined for a space-time path.

Elements	Main abstraction
Space-time path	Fusion of space and time dimensions
States	Changes produced by the connectivity and continuity of a sequence of events
Events	Human activity, the cause for one state change to another, or part of a process
Episodes	Length of time, duration
Evidence	Maps, surveys, satellite images

4.2 THE SPACE-TIME CONFIGURATION

A space-time path can have a longitudinal or branching configuration in which states and events, episodes and evidence are consecutively and chronologically connected through this path. Table 4.1 summarises the main abstractions involved in each of these elements of a space-time path. The origin of a space-time path can be associated with an event element or a state element. For instance, the *creation* of a space-time path can occur by a human action or by the explicit choice of a starting point of interest in an entity state at any particular time. In the STDM both events and states have been modelled as classes of objects. However, they play different roles within the STDM. Events are instances of classes that describe what and when something happened, is happening, or will happen during the lifespans of entities. On the other hand, states are instances of classes that describe what has changed, is changing, or will be changed during the lifespans of entities. The main advantage of this modelling decision is that events can be defined, exist within the database, and interact with the classes of states without depending on the changes in the states themselves.

The longitudinal or branching configuration of a space-time path encapsulates the existence of an entity over space and time. The *existence* will be a sequence of states and events in which change can occur over its states. A change can result from the effects of human activity due to an alteration, modification or transformation of an entity state over time. A space-time path will be an arrangement of events and states that have to be effectively connected to each other to represent the evolution of an entity lifespan. Exploring the temporality and dynamic interactions of classes of events and states may provide an in-depth understanding of our perceptions of the knowledge domain.

The observer's view is a very important factor in determining the spatio-temporal representation. The observer's position on the space-time path will determine their present, hence their past and future. As the observer moves along the space-time path, they experience changes in their perception and understanding of what is past, present and future. The actual notion of present, past and future in the STDM relies on the location of the observer on the space-time path, not the space-time path itself. The historical view of the observer is reduced to the perception of nearness of states and events on the space-time path of a lifespan of an entity.

The demise circumstance characterises the closure of a space-time path. It can occur on the space-time path at any time during the lifespan of an entity. The effectiveness of the STDM in handling the creation, evolution and demise circumstances that can occur on a space-time path is fostered by the instantiation of object classes. An object of a class exists in time and space, whereas a class represents the essence of this object. An object of a class can represent either an event or a state in the STDM.

In the STDM, classes can be categorised as being generic, versioned and unversioned. Each property defined within a class is further classified as being an invariant, version significant or non-version significant attribute. Note how this classification is not deemed to embrace the infinite range of possible changes that can occur over entities of a knowledge domain. The proposed classification is based on changes that can be associated with three update procedures defined in the STDM: (a) creation of a new object, (b) creation of a new object from an existing object, and (c) relocation of an existing object (Table 5.3 gives the complete list of update procedures).

4.3   DATA MODEL CHANGES

The main finding in this investigation is that, although they are primarily concerned with updates in the schema of the spatio-temporal data model, the data model changes also affect the configurations of space-time paths. Data model changes affect the position, orientation and scale of a space-time path in the STDM. This influence has been observed during practical implementation of the STDM in Smallworld GIS. A general overview has evolved from these empirical observations as an attempt to demonstrate how data model changes can affect a space-time path.

In summary the following components of the Banerjee *et al.* classification (see Chapter 3, Section 3, for a detailed description) appear to be important in the present context:

1 Changes to the component of a class
   (a) Changes to properties: they affect the position of a space-time path
   (b) Changes to methods: they affect the orientation of a space-time path
2 Changes to relationships of classes: they affect the orientation of a space-time path
3 Changes to classes themselves: they affect the position and the orientation of a space-time path

A more detailed investigation is usually made into data model changes and how they affect the position, orientation, and scale properties of a space-time path, but it is beyond the scope of this book.

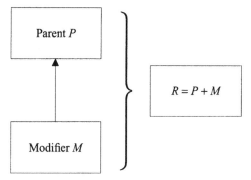

Figure 4.1 Independent incremental modification by inheritance

4.4   CONNECTIVITY ASPECTS

The STDM provides a distinction in version configuration between the system level and the application level. At the system level, version management is supported by object versions that reflect the history of modifications over objects as perceived by the system. At the application level, version management supports the representation of valid time or the sequential development of information as defined by the user or the application.

Considering that states and events from the knowledge domain will be classes of objects within the STDM, the design of an incremental mechanism relies on defining the kind of relationship that should exist among these classes in order to reproduce a space-time path. In fact, this involves an instance of one class being connected with an instance of another class in such a way that it reproduces the history of a particular entity on a space-time path. This has been achieved by using the *inheritance* abstraction to develop an incremental mechanism for the space-time path. Changes happen at the instance level in such a way that an entity has its space-time path depicted by different instances of different classes of objects representing the events and states, which are connected through a mechanism of incremental modification.

Inheritance has been adopted as the incremental modification mechanism for designing space-time paths within the STDM due to its potential for subclassing the evolutionary processes of incremental change of the Time Geography approach. Two types of incremental modification mechanism can be described in the STDM, the independent incremental modification and the overlapping incremental modification.

An *independent incremental modification* by inheritance transforms a parent class P with a modifier M into a resultant class $R = P + M$ (Wegner and Zdonik, 1988). The composition operator $+$ plays an asymmetric role in determining R from P and M (Figure 4.1), since

result R
  inherits P;
  modified by M.

The parent, modifier and result classes exhibit a structure with a finite number of properties:

$$P = (p_1, p_2, \ldots, p_p)$$

$$M = (m_1, m_2, \ldots, m_m)$$

$$R = (r_1, r_2, \ldots, r_r)$$

In the spatio-temporal data model, the independent incremental modification occurs over object classes belonging to the creation, existence and demise circumstances of a space-time path. It represents the connection state–event–state of a space-time path. In this case the parent class is an evolutionary state that is modified by an event (modifier class) into another evolutionary state (resultant class). The properties of the resultant class $(r_1, r_2, \ldots, r_r)$ are independent of those properties of the parent class $(p_1, p_2, \ldots, p_p)$ and the properties of the modifier class $(m_1, m_2, \ldots, m_m)$. The independent incremental mechanism is responsible for the union of these properties in a way that R has $r + p + m$ properties.

Conversely, an *overlapping incremental modification* by inheritance transforms a parent class into a resultant class without having the presence of a modifier class. The properties of the resultant class are inherited from the properties of its parent class (Wegner and Zdonik, 1988). In STDM the overlapping incremental modification occurs over versioned and generic classes. In this case the resultant class is assigned to an instance of the class which inherits some of the properties of its parent class. New properties can be added to the resultant class whose names do not occur in its corresponding parent class.

In conclusion the space-time path is built by connecting each instance of a class with its corresponding instance of another class, according to the incremental modification mechanism involved. This allows us to represent the connection between events and states within the spatio-temporal data model by using the instantiation of object classes and the inheritance relationship between them. Having such a configuration for space-time paths in the spatio-temporal data model means there is a fundamental need for version management of all instances that can belong to a space-time path. This implies the analysis and design of a version management mechanism in order to manage change over time within the spatio-temporal data model.

4.5   VERSION MANAGEMENT

The version management mechanism designed for the STDM implies versioning at the instance level of versioned classes, as well as the classification of their attributes. Versioning at the instance level (object level) uses the available structure of classes in the STDM. In this case, version significant attributes have been implemented by attaching an update method to each one of them. This is achieved by defining trigger update methods for each version significant attribute of these versioned classes. If a user tries to update a version significant attribute, a menu pops up on the screen informing the user to update the version significant attribute in a non-destructive manner. Each version is implemented as a record and methods can be invoked on this record in order to perform the update operations on version significant attributes (see Chapter 3, Section 3.5). The visualisation of versions is achieved by the use of pop-up windows which display a version at the instance level of a versioned class. This is illustrated in the next chapter.

Non-version significant attributes have been incorporated into the STDM in such a way that they are attributes whose values are updated in a destructive manner. Versions are not stored in the database. Invariant attributes have been incorporated into the STDM by attaching an update constraint to the instances of the classes. This is achieved by defining a trigger method for each invariant attribute (record) in a way that its value is not allowed to be updated. This method is triggered every time a user tries to update invariant attributes.

Ordering versions within a spatio-temporal database is also an important aspect in version management. Despite the development of storage devices, such as optical disks, offering new capabilities for storing large amounts of data, a significant waste of storage space will still be minimised by ordering versions compactly. The proper choice of a method for ordering temporal data relies on producing the best performance for a given application. Two general approaches have been developed for databases (Dadam, Lum and Werner, 1984). The first one is the absolute representation for ordering temporal data in which snapshot series are created for each version. The second is the relative approach in which versions are described in relation to a base state that can be located in the present or the past. In both approaches, *version identifiers* are adopted to address versions, and *delta versions* to store a version as a delta version instead of a complete version. Therefore, $Cv_x(n)$ is the version identifier n of the object x, with n_0 being the oldest version and Cv standing for complete version. The delta version is defined as $\Delta_x(k, k')$ where Δ_x is the delta version between the versions number k and k' of an object x. Here are some relevant conclusions on the absolute representation:

- Based on complete versions.

- Versions $(x) = [Cv_x(n), Cv_x(n-1), \ldots, Cv_x(n_0)]$.

- Storage space requirement is prohibitive.

Relative representations are divided into backward and forward. The backward relative representation reconstructs older versions from newer versions:

(a) Backward-oriented
   - Versions $(x) = [Cv_x(n), \Delta_x(n, n-1), \Delta_x(n, n-2), \ldots, \Delta_x(n, n_0)]$.
   - It allows fast access to the current version. Unfortunately, in this versioning strategy, versions have to be recomputed completely whenever a new version is created. This results from the fact that deltas are related to the current version which changes whenever a new version is created.

(b) Backward-oriented accumulative
  - Versions $(x) = [Cv_x(n), \Delta_x(n, n-1), \Delta_x(n-1, n-2), \ldots, \Delta_x(n_{o+1}, n_o)]$.
  - All versions except the current one are expressed as delta to the successor-in-time version.
  - Access time to versions will increase with their age.

The forward relative representation reconstructs newer versions from older versions:

(a) Forward-oriented
  - Versions $(x) = [\Delta_x(n, n_o), \Delta_x(n-1, n_o), \ldots, \Delta_x(n_{o+1}, n_o), Cv_x(n_o)]$.
  - Access time is the same for every version. Each version can be accessed in two steps because all deltas are related to the base version.

(b) Forward-oriented accumulative
  - Versions $(x) = [\Delta_x(n, n-1), \Delta_x(n-1, n-2), \ldots, \Delta_x(n_{o+1}, n_o), Cv_x(n_o)]$.
  - Access time to versions will decrease with their age.

Based on this evaluation by Dadam, Lum and Werner (1984), the backward-oriented accumulative strategy has been adopted in the STDM. The main advantage seen in this strategy is that for each version in the STDM, an accumulative delta is obtained due to a given update. In other words, the delta creation for obtaining the successor-in-time version in the space-time path can be activated by the occurrence of a change within the STDM. Therefore, a successor-in-time version can only be created if an update is triggered within the system. The visualisation of this strategy can be found in the next chapter, where the results are presented.

4.6 CONCLUSIONS

The abstractions developed in the STDM, such as space-time paths incorporating events, constraints and states, can be implemented into GIS. They can be used in modelling a variety of applications in environmental information systems, land information systems and other information systems. One example is the prediction of environmental change due to long-term large-scale climatic variations (Wachowicz and Broadgate, 1993). In this particular knowledge domain, prediction of environmental change requires an understanding of the principal mechanisms implicated in long-term large-scale climatic variation. Uncovering these mechanisms can only be achieved by analysing past environment states as well as recognising patterns of change through time. The STDM provides the semantics of events and states that can be used to describe the environmental changes. They can also be categorised as effect (an environmental change which can be detected by experience or observation of the environment) and cause (circumstances acting over a period of time which produce an environmental change). The STDM proposed in this book enables the time geographic framing of past environmental states (changes) by exploring the derivation of events from these states, and vice versa, due to their spatio-temporal interdependence.

Evaluating a spatio-temporal data model is a difficult and drawn-out process. The approach taken in this book has been to construct a prototype implementation as complete as possible within the time and resources available and to use it for evaluating the feasibility of the STDM, along with its strengths and weaknesses. This is discussed in more detail in the next chapter. A complete evaluation, in particular an implementation of a whole application system, would take much more time than was available.

The results of the implementation are presented in the next chapter and show that the prototype implementation preserves the principal features of the STDM. Care had to be taken when implementing these features, such as space-time paths and their associated incremental modification mechanism, to preserve the functionality of the spatio-temporal data model. The resulting implementation is therefore well placed to take full advantage of the object-oriented constructs (object identifier, inheritance, polymorphism).

The strengths of the STDM and its implementation lie in the well-defined 'object representation' that has been developed. This representation offers much practical guidance, and implies the feasibility of applying object-oriented methods to the problem of handling space and time in GIS. The Time Geography approach brings to light the link between components as events, states and constraints, and also much of the functionality of modelling space and time in the context of the knowledge domains in GIS.

The weaknesses lie in the lack of an appropriate access method for implementing space-time paths. New indexing techniques need to be developed in order to avoid problems in the implementation of these paths. This problem is not addressed here, but many new indexing techniques have been proposed in the literature (Schneider and Kriegel, 1992; Renolen, 1996).

Although this research concentrates on spatio-temporal data modelling, it is hoped the findings will contribute more widely to the next generation of geographic information systems, with their improved capabilities for handling spatio-temporal data. The STDM presented in this book emphasises many desirable characteristics for a temporal GIS:

- A continuous space-time path that supports modelling states and events in order to integrate space and time in a GIS.
- A sensible compromise between the flexibility offered by object-oriented methods and the drawbacks of implementing an object-oriented data model in a GIS.
- Scope for handling changes within a GIS.

But further research is necessary, especially in these three areas:

- To develop new indexing techniques for spatio-temporal data.
- To investigate new approaches for storing and manipulating topological relationships in a GIS, in order to avoid adding unnecessary complexity to the spatio-temporal data model.
- To develop temporal capabilities such as dynamic visualisation and spatio-temporal queries.

CHAPTER FIVE

Applying the STDM:
public boundaries evolution

The act of drawing a diagram does not constitute analysis or design. A diagram simply captures a statement of a system's behaviour (for analysis), or the vision and details of an architecture (for design). If you follow the work of any engineer – software, civil, mechanical, chemical, architectural, or whatever – you will soon realize that the one and only place that a system is conceived is in the mind of the designer. As this design unfolds over time, it is often captured on such high-tech media as white boards, napkins, and the backs of envelopes.

G. Booch, *Object-Oriented Analysis and Design with Applications*

This chapter provides a comprehensive set of diagrams to illustrate the significant modelling decisions when applying the STDM to a knowledge domain. And the knowledge domain has been chosen from historical geography – maintenance of the political boundary record. It is illustrated by a process diagram that describes a set of processes for four different scenarios and their allocation to processors in the physical view of the STDM. Class diagrams describe the relationships among states and events, as well as the operations and properties associated with them. They are also used to describe the semantic dependency between classes and the ability to walk through the model from one scenario to another. Finally, interaction diagrams illustrate the execution of each scenario and can be used to visualise the process involved in 'making space-time paths' within the STDM. ObjectMaker release 2.1 by Mark V Systems Limited has been used as the CASE tool for supporting the proposed diagrams of the STDM. The key to the symbols is given in Appendix A.

5.1 PUBLIC BOUNDARY RECORD MAINTENANCE

Public boundaries represent the line of physical contact between administrative units in Great Britain. The arrangement of public boundaries forms an irregular tessellation of polygons that represent the whole hierarchy of local government and

European constituency areas. Approximately 3000 changes occur to public boundaries in England every year, increasing the volume of data at the rate of 5 to 10 megabytes a year (Rackham, 1992). Records of the history of each public boundary can be found in documents and published material such as parliamentary acts and orders retained by the Ordnance Survey. These contain legal information defining the line on the basis of which each public boundary is attached to a physical feature on the ground.

Maps portraying the public boundaries when they have been delimited and demarcated also contain important historical information for identifying the physical features on the ground. Fieldwork records kept by the Ordnance Survey are another valuable source of historical information about the evolution of public boundaries. One example is the register of discrepancies between the true (legal) description of a public boundary and its demarcation on the ground. Another example is the collection of perambulation cards noting the occurrence of some change in the location of a physical feature on the ground from its previous public boundary demarcation.

5.1.1   The knowledge domain

Public boundaries are linear objects that experience a succession of changes in their positions during their lifespans. The history of each public boundary is unique and shows the geographical significance of a public boundary over the development of landscapes, socio-economic policies and historical conflicts. Attempts to devise historical processes involved in political boundary evolution have been successful in identifying a set of reliable events by which human actions can be connected with the evolution of the majority of public boundaries. De Lapradelle (1928) identified preparation, decision and execution as the three main historical events which most political boundaries go through:

> The [event] . . . of preparation precedes true delimitation. The problem of the boundary's location is debated first at the political level then at the technical level. The question is, in general, of determining, without complete territorial debate, the principal alignment which the boundary will follow. . . . The decision involves the description of the boundary or delimitation. . . . The execution consists of marking on the ground the boundary which has been described and adopted, an operation which carries the name demarcation.   (de Lapradelle, 1928, p. 73)

In adopting these first delineations for historical events in political boundary evolution, Jones (1945) extended them to allocation, delimitation, demarcation and administration. The administration event would deal with the maintenance of the physical features which have been allocated to be a public boundary. More recently, Prescott (1987) pointed out the following three main processes in political boundary evolution:

- *evolution in definition*: The historical events suggested by Lapradelle and Jones are proposed as boundary-making events; 'allocation refers to the political decision on the distribution of territory; delimitation involves the selection of a specific boundary site; demarcation concerns the marking of the boundary on the ground; and administration relates to the provisions for supervising the maintenance of the boundary' (*ibid.*, p. 69).

- *evolution in position*: This means 'how long the boundary has occupied particular sites' (*ibid.*, p. 77).

- *evolution in the state functions*: Evolution of state functions applied at the boundary means 'the effectiveness with which the boundary marks the limits of sovereignty' (*ibid.*, p. 80).

Analysing the evolutionary aspects of political boundaries represents the study of human activities that have been relevant to the location of a particular boundary. The STDM considered here was formulated and constructed using evolution in definition. This process is characterised by the changes related to different states acquired by every individual boundary after going through its historical events (allocation, delimitation, demarcation, administration).

Rackham (1987, pp. 32–3) identified from Booth (1980) six different states which most public boundaries can go through:

- *Draft*: proposed but not yet confirmed by an act or order.

- *Proposed works*: referred to in an order related to a physical feature which has not yet been constructed (e.g. a new road).

- *New*: made in an act or order but not yet mered.

- *Disputed*: mered but not certified because of some disagreement.

- *Old*: ascertained on the ground, certified by the relevant authorities and therefore fixed in alignment.

- *Obsolete*: old boundary no longer used to demarcate administrative areas (obsolete boundary may revert to old boundary if it is reused at a later date to demarcate an administrative area).

The main challenge in designing a spatio-temporal data model for this knowledge domain is to handle systematically the changes related to different *states* (draft, proposed work, new, disputed, old, obsolete) acquired by every public boundary after going through its historical *events* (allocation, delimitation, demarcation, administration). This can be achieved by exploring the synergy of historical events and states of public boundaries through the creation of space-time paths for each public boundary.

5.1.2 The space-time path

Each public boundary has its own space-time path that represents its lifespan. As mentioned in Chapter 4, the STDM is deemed to deal with the explanatory task in the spatio-temporal reasoning domain. This involves producing a description of the

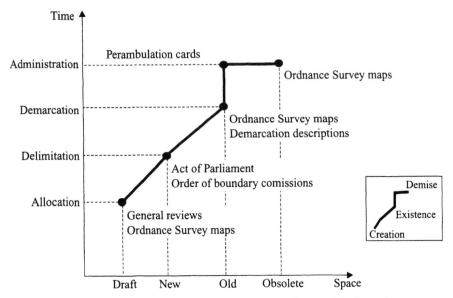

Figure 5.1 An example of a possible space-time path for a public boundary

evolution of public boundaries at an earlier time that accounts for the public boundaries being the way they are at a later time. Considering the evolution of public
boundaries, this explanatory task implies that the space-time path has a longitudinal
configuration. There is no possibility of having a branching configuration for a
space-time path of a public boundary over its past. The creation, existence and
demise circumstances are primary components of a space-time path.

Creation

Creation represents the space-time origin of a public boundary lifespan that will
evolve towards the future or past from its origin (Figure 5.1). At the creation
circumstance, an *allocation* event takes place in selecting a ground feature to be a
future public boundary. Allocation events are responsible for representing the dimension valid time within the STDM. Time is represented as a nominal value that
indicates the actual date when the allocation took place. One example of this case is
the general reviews of public boundaries carried out by parliamentary boundary
commissions and local government commissions (Coombes *et al.*, 1993). These
reviews are needed to investigate positions or arrangements of new public boundaries to ensure a uniform representation of the electoral population for every constituency (ward, electoral division, district, region and parliamentary constituency)
in Great Britain. The parliamentary boundary commissions have carried out three
general reviews since 1944. If the creation circumstance is set on the space-time
path for occurring in 1994, the space-time path will deal with the task of modelling
the boundary changes as investigated by the commissioners since 1944. Once the

creation circumstance is dated, it cannot subsequently be modified since this would cause integrity problems within the STDM.

There exists a specific spatial relationship between a ground feature and a public boundary that has to be selected from a set of possibilities depending on the kind of ground feature being utilised. For example, ground features can be paths, ponds, rivers, railways, fences, roads and hedges. Therefore, some possible spatial relationships would be 'centre of' the road, 'face of' the fence, 'root of' the hedge, '1.00 m from' the railway or '1.83 m from' the path. Public boundaries have to be related to ground features, but, some landscapes do not have suitable ground features. In this case, a straight line between two mereing points on the ground determines a public boundary. Otherwise, an engineering work can be carried out to build the necessary ground feature to be a part of the public boundary. All boundary lines representing these spatial relationships are portrayed on Ordnance Survey maps at 1 : 1250, 1 : 2500 and 1 : 10 000 scales.

Existence

Existence encapsulates the space-time path over which the historical *events* (allocation, delimitation, demarcation, administration) of a public boundary evolve over space and time. The existence of a boundary is also constituted by the occurrence of changes. These changes result from the effects of human activity. They are the *states* through which the public boundaries evolve. For example, a draft state is assigned to a public boundary which has not yet been confirmed by an act of Parliament. Once this act is promulgated, the boundary modifies its *draft* state to a different state named as *new*. The historical event to occur is the *delimitation* which generates the *new* boundary state (see Figure 5.1).

The operative date and the effective date represent the episode element of the space-time path for the delimitation event. An operative date is the actual date when a public boundary is issued by an act or order, generally 16 May, or the first Thursday in May for Scotland. An effective date assigns the date when an act or order has become effectual after the General Election following the operative date (Rackham, 1987). A coupling constraint for each public boundary is essential to guarantee that the date in the allocation event must be prior to both operative and effective dates of the delimitation event (Chapter 2 describes coupling, capability and authority constraints).

All public boundaries in Great Britain have been delimited by the issuing of an act of Parliament or an order of the boundary commission. Extensive archives containing maps at 1 : 10 000 scale and the statutory documents, such as acts and orders, are held by the Ordnance Survey in order to preserve the legal records of the public boundaries. Thus, the new boundary state plays an important role in the STDM, which verifies the fact that each public boundary cannot exist without having a new boundary state (coupling constraint).

The significant characteristic encountered in the spatial relationship between draft and new states is spatial generalisation, which demands procedures for line simplification (capability constraint). The Ordnance Survey uses different scales for portraying a public boundary having draft and new states. A new boundary is usually portrayed at larger scales than a draft boundary. As a result, some points have to be eliminated from the set of points for the draft boundary line. However, turning points might have to be preserved as intact points representing the topological junctions, i.e. the line intersections between public boundaries.

At this point of the space-time path, a political boundary can be demarcated on the ground, thus the *demarcation* event occurs and the *old* state is set on the space-time path (see Figure 5.1). All old boundaries are portrayed on basic maps (1 : 1250, 1 : 2500 and 1 : 10 000 scales) by boundary lines and by symbols representing their respective demarcation descriptions. The spatial relationship between a new boundary state and an old boundary state plays an important role in detecting any controversy between the interpretation of the legal definition of a public boundary and its equivalent geographical position on the landscape. Sometimes this controversy can provoke boundary disputes over the actual location of a public boundary. This dichotomy can arise for several reasons, such as having more than one interpretation of terms used in the delimitation event, as well as having a contradictory demarcation of the turning points along the boundary line. Generally, the uncertainty of geographical interpretation is more likely to be the culprit.

The episode for the demarcation event has an interval representation that is required to date the start and end of the demarcation event for a public boundary. Since most of the disputes concerning the actual location of a public boundary occur during the demarcation events, it is fundamental to have the dates when the dispute began, as well as the dates when actions were taken to rectify the disagreement.

Finally, the administration event can take place in a space-time path. The main reason for having *administration* events within a space-time path is the fact that a public boundary can change its position on the ground (see Figure 5.1). As the boundary changes its position, a transfer of territory from one authority to another will occur, causing changes in sovereignty and, possibly, changes in the socio-economic development of the border landscapes. The updates on a position of a boundary can occur by three basic changes:

- *Natural changes* The most common example is the displacement of a boundary position with the displacement of the watercourse by rivers and streams.

- *Man-made alterations* Updates are due to changes in the position of a boundary by opencast mining, erosion or overthrow of the ground features.

- *Attachment* Attachment of new descriptions to an existing boundary can lead to its position being updated.

New descriptions can also occur at any time in the existence of a public boundary. However, they are more likely to appear during the delimitation event when a boundary line is incorrectly portrayed on the original map in relation to its true position on the ground, and much later when the position of a boundary has been incorrectly demarcated on the ground.

The Ordnance Survey maintains regular perambulation measurements by which the surveyor confirms the displacement of the ground feature on the landscape. This survey plays an important role in the space-time path of the STDM since it represents the temporal relationship between the moment when the ground feature is updated on the map and the moment when the equivalent change is confirmed in the landscape. On the other hand, perambulation measurements also uncover mistakes and misinterpretations of the geographical terms used in the delimitation events. A common example is that a political boundary can be correctly portrayed on a map, but the actual boundary line was wrongly demarcated on the landscape.

Demise

Demise characterises the closure of a space-time path. It can occur on the space-time path at any time during the lifespan of a public boundary. However, this is more likely to occur when a public boundary is no longer operative or effective. In other words, when a public boundary reaches its obsolete state (see Figure 5.1).

5.2 EVOLUTION IN DEFINITION

The events and states from the space-time paths have been modelled as object classes. However, they play different roles within the STDM. Events are used to describe what happened, is happening, or will happen during the lifespans of public boundaries. On the other hand, states tell us what has changed, is changing, or will be changed during the lifespan of public boundaries. The main advantage of this modelling decision is that events can be modelled, exist within the database, and interact with the states of a public boundary without depending on the changes in the states themselves. Changes occur at the instance level (object level) in such a way that a public boundary has its space-time path depicted by different instances of different classes representing the events and states. All the instances are connected through the incremental modification mechanism based on the inheritance construct of object orientation (see Chapter 4, Section 4).

`GroundFeature` and `PublicBoundary` have been identified as generic classes within the STDM (Chapter 3, Section 3, explains system-defined classes). Each of them contains objects which embody some state, exhibit certain behaviour, and are uniquely identifiable. `GroundFeature` represents every physical feature in the landscape that has been assigned to be a public boundary object. Likewise, the `PublicBoundary` class denotes the political boundary itself. Considering the generic classes of the STDM, we can now associate them with their respective system-defined types in the following manner:

- *Generic classes*
  `PublicBoundary`
  `GroundFeature`

- *Versioned classes*
  `DraftBoundary`   `ObsoleteBoundary`
  `NewBoundary`   `GroundFeatureRevolutionaryState`
  `OldBoundary`   `OldBoundaryRevolutionaryState`

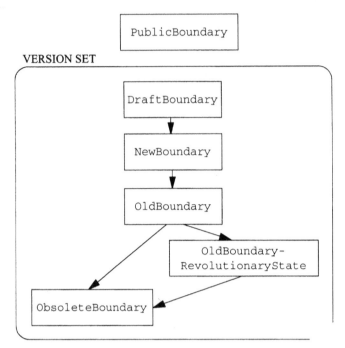

Figure 5.2   The version graph of the STDM

- *Unversioned classes*
  Assumption   Demarcation
  Allocation   Perambulation
  Delimitation

Versioned classes represent the different evolutionary states within the STDM. In essence the lifespan of a public boundary can be represented by these versioned classes. The instances of the versioned classes are the possible states of the space-time path of a political boundary. In other words, each state of a public boundary belongs to a different versioned class in order to create space-time paths within the system. Figure 5.2 illustrates the version graph that represents the link between versioned classes that contain the versions themselves in the STDM.

A version graph plays an important role in the temporal data management of the STDM because it helps to visualise the space-time path without depicting the events. Therefore it can be used as a modelling tool for designing the version management mechanism within the system. In defining what versioned classes are needed in the STDM, the next step is related to how the instances of these classes (i.e. the versions) should be identified and distinguished within the STDM. This is discussed in the following sections.

Unversioned classes represent the events over the space-time path, therefore they hold a time stamp which is the valid time corresponding to the lifespan of the state of a public boundary or the occurrence of an event. In the STDM, the time stamp is an attribute value associated with the valid time that can be nominal type (e.g. 27 May) or ratio type (e.g. event demarcation starts on 17 June and ends on 29 July).

The next section considers the four main scenarios devised as integrated subsystems within the STDM:

- *Public boundary entry scenario* Based on the creation circumstance of the spacetime path, it is responsible for managing the allocation events in which a ground feature is assigned to be a public boundary.

- *Evolution tracking scenario* Based on the existence circumstance of the spacetime path, it is in charge of managing all possible states of a public boundary (draft, proposed work, new, disputed, old, obsolete) and their respective historical events (allocation, delimitation, demarcation).

- *Update scenario* Based on the existence circumstance of the space-time path, it is responsible for updating public boundaries and managing all changes in the position of a public boundary.

- *Archiving scenario* Based on the demise circumstance of the space-time path, it is in charge of storing and retrieving the obsolete public boundaries.

5.3 FOUR MAIN SCENARIOS

Each of the four scenarios has been allocated differently in the physical view of the system. The process diagram (Figure 5.3) illustrates the different scenarios by

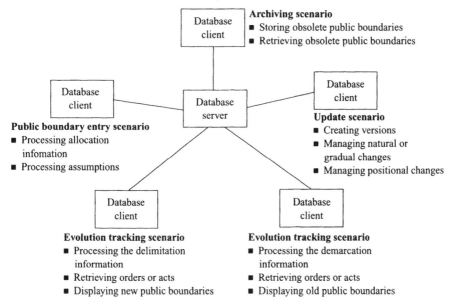

Figure 5.3 The process diagram

allocating the appropriate processes to the processors in the physical view of the STDM. It can also be used to visualise the mapping of the architecture onto its realisation in code, and it shows the overlap between the analysis and design activities in developing the spatio-temporal model. Each scenario represents a potential path area at the analysis level as well as a client–server architecture of the system at the design level. This suggests a decentralised architecture, so most of the processing can be done by the client, reducing the bottleneck across the network.

5.3.1   Public boundary entry scenario

The public boundary entry scenario represents the creation circumstance of the space-time path within the STDM. It manages the states and events concerned with the allocation events in the STDM. And it involves the creation of the space-time path itself within the model. Four classes have been designed for the public boundary entry scenario: DraftBoundary, GroundFeature, Assumption and Allocation. The GroundFeature class represents the ground features on the landscape about which statements might be made. The DraftBoundary class comprises many GroundFeature objects, each of which might have different statements about them. The statements play an important role in the public boundary entry scenario. A statement can be related to an assumption which states that a ground feature can be regarded as a possible feature to be a public boundary. This denotes the mapping between an instance of the GroundFeature class and its corresponding instance of the Assumption class.

Once a ground feature has been selected to be a public boundary, this denotes the mapping between an instance of the GroundFeature class and its corresponding instance of the Allocation class. In this case the mapping is definite, hence unchangeable, because it depicts the creation of a space-time path within the system. As a result, an instance of the DraftBoundary class can be created. The class diagram in Figure 5.4 illustrates the design decisions regarding these object classes (Appendix C gives the entire class diagram of the STDM).

An inheritance relationship exists between the Allocation and Assumption classes. The Assumption class represents a superclass having the generalised statements about the allocation event. The Allocation class is a subclass representing a specialisation in which are added properties and methods from the superclass Assumption.

The aggregation relationship is assigned between the Allocation class and the GroundFeature class. This abstraction permits different instances of the GroundFeature class to be allocated as a possible public boundary. Allocating ground features to be a public boundary is deemed to occur in such a way that selecting a ground feature to be an instance of the GroundFeature class does not interfere with the properties of its corresponding instance of the Allocation class as a whole. And removing an instance of the GroundFeature class does not necessarily delete all its corresponding instances of the Allocation class.

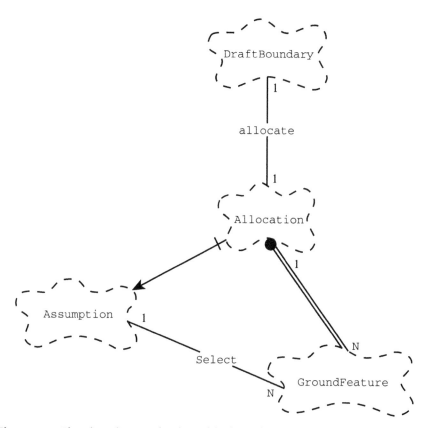

Figure 5.4 The class diagram for the public boundary entry scenario

The allocate and select associations in the class diagram denote a semantic dependency and suggest a bidirectional navigation between classes. For example, given an instance of GroundFeature, we should be able to locate the respective object denoting the DraftBoundary, and vice versa. Besides, these associations represent the independent incremental modifications in the STDM.

In this case, GroundFeature is the parent class having Allocation and Assumption as modifiers into a resultant DraftBoundary class. Table 5.1 shows the properties defined for each of the classes involved in the public boundary entry scenario. Although these classes are interrelated through an incremental mechanism, each one has its own properties. The HistoricalView class in the public boundary entry scenario represents this independent incremental mechanism of the STDM. It denotes the union of all properties which belong to the parent, modifier and resultant classes in the scenario (Chapter 4, Section 4 explains the independent incremental modification of space-time paths). As a design decision, the HistoricalView class has been defined for visualising the space-time path involved in the public boundary entry scenario. The historical view provides a snapshot of all the properties within the system, which have been involved in the

Table 5.1 Class properties.

Class	Properties
DraftBoundary	line
	length
	mereing_point[a]
	type {*draft creation not confirmed, draft creation confirmed, proposed works*}
	description {*base of, centre of, foot of, 1.00 m from, not applicable*}
GroundFeature	point
	line
	area
	type_ground_feature (p. 102) {*stream, fence, hedge, wall, pond, undefined, defaced*}
Allocation	map_scale_used_for
	description
	date [*year, month, day*]
Assumption	statement
	validated {*yes, no*}
	valid [*from, to*]

[a] See Appendix B.

assumption and allocation events of a specific public boundary selected by the user up to a certain point in time.

During the execution of the public boundary entry scenario, many statements mapped through the Assumption class will be made about the feasibility of certain ground features becoming a public boundary. As the scenario moves closer to a final decision, these statements eventually become Allocation objects. The execution of this scenario is illustrated in Figure 5.5; this interaction diagram provides a global perspective of the various operations involved in the public boundary entry scenario, and it shows the behaviour of the system in terms of the interaction between instances of classes. These are the operations designated for the following classes:

- GroundFeature
  draftBoundary
- DraftBoundary
  groundFeature
- Assumption
  make   select
  value   become
  mostRecent
- Allocation
  select
  assign

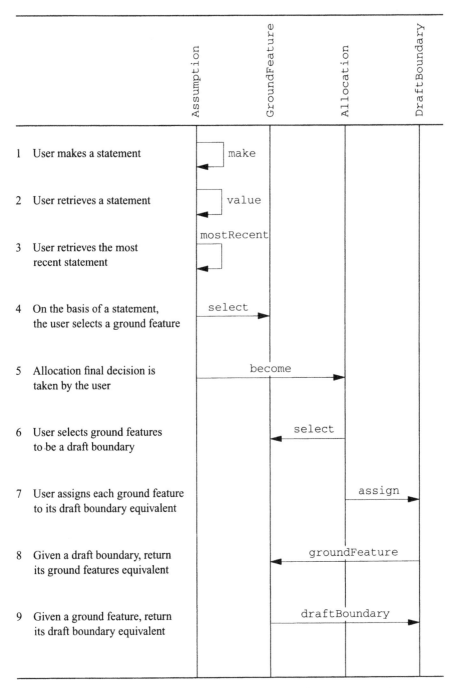

Figure 5.5 The interaction diagram for the public boundary entry scenario

5.3.2 Evolution tracking scenario

The main role of the evolution tracking scenario is to assign an order among events as well as the precise dates for the boundary-making process. Therefore, both delimitation and demarcation events from the space-time path are described in the evolution tracking scenario (Appendix C gives the entire class diagram of the model). In the delimitation event, an act or order confirms a draft boundary as a public boundary, so the boundary assumes a new evolutionary state in the STDM, called a new boundary state. The focus is on the design decisions regarding the structure and behaviour presented by the instances of the classes DraftBoundary, Delimitation and NewBoundary.

For the demarcation event, the position of a new boundary is ratified on the ground by surveyors, on the basis of the delimitation documents. In the STDM, the emphasis is on modelling the controversial aspects between the delimitation and demarcation of a public boundary over time. This involves the classes NewBoundary, Demarcation, and OldBoundary.

The Delimitation and Demarcation classes characterise the delimitation and demarcation events. NewBoundary and OldBoundary are classes defined to represent the new and old evolutionary states of a public boundary. On the basis of the analysis carried out for modelling the delimitation events, a prerequisite arises to carry out a line generalisation of a public boundary. This imposes different scales for portraying the same line that belongs to the Draft-Boundary and later on when it has been assigned to be a NewBoundary. For example, consider displaying a specific line of a public boundary having a draft status at 1 : 2500 scale. Once this line has been elected to be a new boundary, its display might appear at one unique scale, i.e. at 1 : 10 000 scale. And for clarity, some previous points belonging to the line in its draft state have to be eliminated.

There are two ways to produce a visual representation of the different states of a public boundary. An external class can be defined in order to query each object in terms of each kind of graphical display to be employed. Alternatively, each object can encapsulate the knowledge of how to display itself. The external class has been chosen as the better solution due to its closeness to the object-oriented concepts. For example, an object will be a line representing the public boundary. Having a draft state, this line will be displayed at 1 : 1250, 1 : 2500 or 1 : 10 000 scale with the following graphical elements (Appendix B):

- Points portraying the change of a ground feature description (mereing points).
- Text properties portraying the description of the line, e.g. CR (centre of road), CS (centre of stream) and FF (face of fence).

In contrast, by having a new state, the same line will be displayed at 1 : 10 000 scale with the following graphical elements (Appendix B):

- Turning points portraying the line intersections between public boundaries.

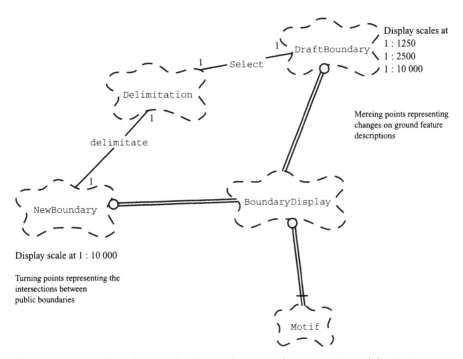

Figure 5.6 The class diagram for the evolution tracking scenario – delimitation event

Figure 5.6 illustrates the DraftBoundary and NewBoundary classes sharing a common class BoundaryDisplay through a *using* relationship within the model. Both DraftBoundary and NewBoundary are the clients of the display inter- face of the supplier BoundaryDisplay. The BoundaryDisplay provides the display routines for the graphical elements that each client requires. In implementa- tion terms, the BoundaryDisplay is built upon a Motif class which deals with the off-the-shelf graphics package. The advantage of this design is that it allows a future replacement of the available display software with, for example, a hypermedia coordination which would display the objects in a dynamic manner. This would require the replacement of the display routines in the BoundaryDisplay class without the need to modify the implementation of every displayable object of DraftBoundary and NewBoundary.

The execution of this scenario is illustrated in Figure 5.7. The interaction dia- gram illustrates the behaviour of the system in terms of the interaction between instances of classes. The following operations have been identified for the respect- ive classes given below:

- DraftBoundary
    select
    notify
    length

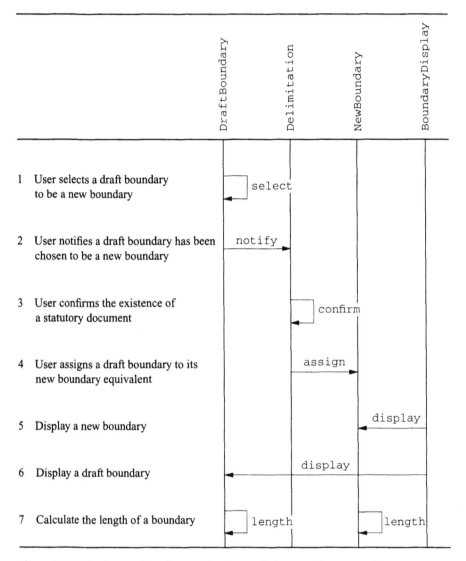

1   User selects a draft boundary
    to be a new boundary

2   User notifies a draft boundary has been
    chosen to be a new boundary

3   User confirms the existence of
    a statutory document

4   User assigns a draft boundary to its
    new boundary equivalent

5   Display a new boundary

6   Display a draft boundary

7   Calculate the length of a boundary

Figure 5.7   The interaction diagram for the evolution tracking scenario –
delimitation event

- Delimitation
    confirm
    assign
- NewBoundary
    length
- BoundaryDisplay
    display

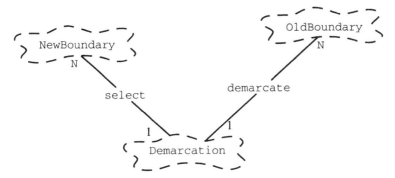

Figure 5.8 The class diagram for the evolution tracking scenario – demarcation event

In the evolution tracking scenario, the demarcation event involves the NewBoundary class, the Demarcation class and the OldBoundary class (Figure 5.8). NewBoundary represents the legal location of a public boundary. OldBoundary represents the old status of a public boundary. In other words, it characterises the actual location of a public boundary after the demarcation. The Demarcation class plays an important role within the STDM. It controls the flow of operations between NewBoundary and OldBoundary.

The interaction diagram in Figure 5.9 uses a box to represent the relative time that the flow of control is focused in a Demarcation class. Each instance of Demarcation is the ultimate focus of control, and its behaviour of carrying out a demarcation event invokes different methods over the instances of NewBoundary and OldBoundary. This is achieved by defining the following operations for the Demarcation class: select, confirm, assign.

The delimitate, demarcate and select associations in the class diagrams (Figures 5.6 and 5.8) denote a semantic dependency and suggest a bidirectional navigation between classes. They also represent the independent incremental modifications in the STDM. In the delimitation event, DraftBoundary is the parent class having Delimitation as the modifier into a resultant NewBoundary class. Moreover, in the demarcation event, NewBoundary becomes the parent class having Demarcation as the modifier into a resultant OldBoundary class.

Consequently, a user will be able to select a draft boundary to be a new boundary, and afterwards, to be an old boundary using the independent incremental mechanism. This allows users to enter values for the properties of classes involved in the evolution tracking scenario, taking into account the evolutionary aspects of the scenario. Classes are interrelated through the incremental mechanism in order to support space-time paths. Their properties do not depend on the existence of such a liaison between the classes. The independent incremental modification ensures that each of these classes has an independent lifespan. HistoricalView classes have been defined for visualising the space-time paths in the evolution tracking scenario, and Table 5.2 shows the properties attached to some of the other classes.

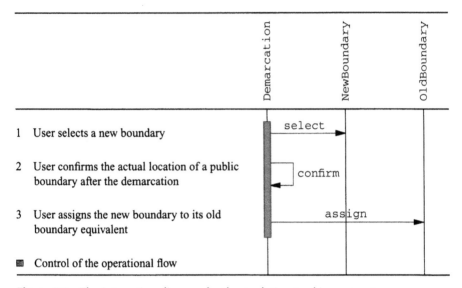

Figure 5.9 The interaction diagram for the evolution tracking scenario –
demarcation event

Table 5.2 Class properties.

Class	Properties

Delimitation	statutory_document
	      {*act of parliament*
	            boundary commission order}
	map_scale_used_for
	operative_date [*year, month, day*]
	effective_date [*year, month, day*]
	actual_date [*year, month, day*]

Demarcation	surveyor_name map_scale_used_for valid {*from, to*}

NewBoundary	line
	turning-point
	length
	description {*operative*[a]
	            operative not effective[b]
	            operative effective[c]}

OldBoundary	line turning-point length type {*old, current, disputed*}

[a] Assigned by an act or order.
[b] Assigned by an act or order which has not yet become effective.
[c] Assigned by an act or order already effective.

Table 5.3 Vector representation: main update procedures.

Update procedure	Change
1 Creation of a new object	nothing
2 Creation of a new object from an existing object	geometry, topology, theme
3 Status updating of an object	theme
4 Relocation of an existing object	geometry
5 Alteration of the spatial relationships among objects	topology
6 Relocation of an existing object with an alteration of its spatial relationship with other objects	geometry, topology
7 Status updating of an object with alteration of its spatial relationship with other objects	theme, topology
8 Status updating and relocation of an existing object	theme, geometry

5.3.3 Update scenario

In the update scenario every change is due to an update procedure, but not every update causes a change of state. When an update procedure generates changes, they are generated according to the spatial data representation being used in the GIS context. Among the spatial data representations available within GIS, the vector representation is the most complete because the geometric, topological and thematic properties of an object class can all be employed to describe changes. In contrast, a grid cell (raster) representation allows changes to be described using only thematic characteristics. For a vector representation, Armstrong (1988) defines eight possible update procedures and the changes that accompany them (Table 5.3).

The update scenario is deemed to handle the changes due to natural changes and new demarcation descriptions to public boundaries. Natural changes can be linked to update procedures 2, 4, 5 and 6 (Table 5.3). On the other hand, new demarcation descriptions in the update scenario are associated with update procedures 1, 2 and 4. So update procedures 1, 2 and 4 have been taken to illustrate the update scenario of a public boundary within the STDM. These update procedures have been designed to operate over the GroundFeature and OldBoundary classes in such a way that states are created for a public boundary. The class diagram in Figure 5.10 illustrates the design decisions regarding these object classes (Appendix C gives the entire class diagram of the STDM).

Both states on the diagram present an inheritance relationship with their corresponding instances. For example, an instance of the OldBoundary-RevolutionaryState class is created by inheriting some properties from its corresponding OldBoundary class. In this case the STDM is deemed to manage the overlapping incremental modification. New properties can be added to an instance of the OldBoundaryRevolutionaryState class whose names do not occur in its corresponding instance of the OldBoundary class. Table 5.4

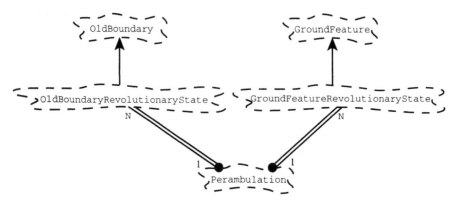

Figure 5.10 The class diagram for the update scenario

Table 5.4 Properties of GroundFeatureRevolutionaryState and OldBoundaryRevolutionaryState classes.

Class	Properties
GroundFeatureRevolutionaryState	type[a]_ground_feature updated_point updated_line updated_area
OldBoundaryRevolutionaryState	type[a] updated_line updated_turning_point length

[a] Inherited properties from their respective parent classes.

illustrates the properties of the GroundFeatureRevolutionaryState class and the OldBoundaryRevolutionaryState class.

The exploratory nature of this design embraces two main aspects in creating discrete states using the overlapping incremental modification. First, the instances of the GroundFeature and OldBoundary classes are subject to the update procedures previously described. Second, there can exist several states for the same GroundFeature and OldBoundary classes. The update procedures play an important role in the update scenario. Figure 5.11 illustrates an example with three update procedures: creation of a new object, creation of a new object from an existing object, and relocation of an existing object.

5.3.4 Archiving scenario

Two classes have been defined in this scenario: OldBoundary represents the old state of a public boundary and ObsoleteBoundary represents the obsolete state

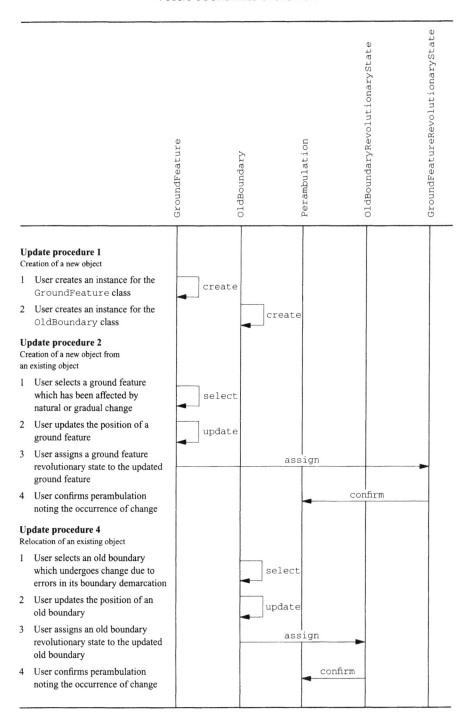

Figure 5.11 The interaction diagram for the update scenario

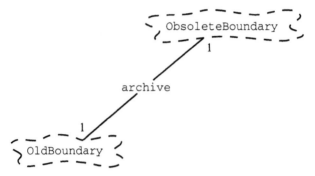

Figure 5.12   The class diagram for the archiving scenario

of a public boundary (Figure 5.12). This assumes the current data are associated to the `OldBoundary` class. Meanwhile `ObsoleteBoundary` represents the historical data. Current data are likely to have a higher query access frequency. Historical data archives will tend to become larger, hence they are most likely to be stored on cheaper optical disk. The execution of this scenario can be illustrated on an interaction diagram (Figure 5.13). The trigger method append assumes the current data are stored separately from the historical data, each kind having its own access method and possibly its own storage medium.

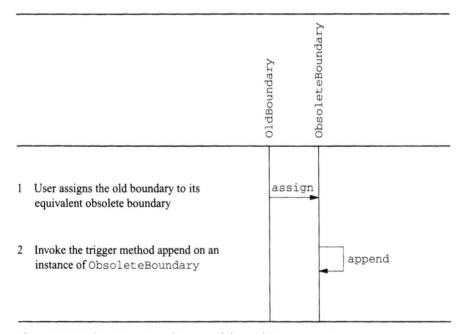

Figure 5.13   The interaction diagram of the archiving scenario

Table 5.5 Versioned and generic classes: classification of attributes.

DraftBoundary

Version significant	none
Non-version significant	line, length, mereing_point, type, description
Invariant	none

NewBoundary

Version significant	none
Non-version significant	line, turning_point, length, description
Invariant	none

OldBoundary

Version significant	line
Non-version significant	none
Invariant	type

ObsoleteBoundary

Version significant	none
Non-version significant	none
Invariant	line, turning_point, length, type

OldBoundaryRevolutionaryState

Version significant	none
Non-version significant	none
Invariant	updated_line, type

GroundFeature

Version significant	point, line, area
Non-version significant	none
Invariant	type_ground_feature

GroundFeatureRevolutionaryState

Version significant	none
Non-version significant	none
Invariant	updated_point, updated_line, updated_area, type_ground_feature

5.4 VERSION MANAGEMENT

A different identifier (OID) has been associated with each version (state) in the STDM. This strategy has been taken in the STDM because it emerges as a more suitable way to track versions. The incremental modification approach designed in the STDM is based on an inheritance mechanism for the properties of an object, so it would be unfeasible to attach the versions at the attribute level. Versioning needs to be done at the instance level, i.e. at the object level. The properties of the generic and versioned classes of the STDM have therefore been divided into the three main categories described in Chapter 3 (invariant, version significant, non-version significant).

This classification is not deemed to embrace the infinite range of valid changes which can occur over public boundaries. It is based on the valid changes which have been associated with the update procedures defined for the STDM. They are deemed to handle the valid changes due to natural changes and due to new demarcation descriptions of public boundaries (see Chapter 3). Table 5.5 illustrates the classification, bearing in mind the update procedures previously defined in the STDM: creation of a new object, creation of a new object from an existing object, and relocation of an existing object. These update procedures are due to valid changes that have occurred over a public boundary. In the STDM they are applied to the `GroundFeature` and `OldBoundary` classes. Consequently, the classification shows both classes having version significant attributes. These version significant attributes must be updated in a structured manner. An update in one of these attributes pertaining to the `GroundFeature` class and the `OldBoundary` class creates versions which are respectively the instances of the `GroundFeatureRevolutionaryState` class and the `OldBoundaryRevolutionaryState` class (see Figure 5.10).

This classification has provided a unifying solution for effective management of versions within the STDM. It means that maintenance of versions within the STDM can be implemented using a GIS. Hence the abstractions developed in the STDM, such as space-time paths involving events and states, can also be implemented using a GIS. Version management will provide a mechanism for maintaining consistent update propagation that prevents unnecessary version proliferation within the system. And beyond the evolution of public boundaries, it may have much to offer other applications, both conceptually and practically.

5.5 CONCLUSIONS

Developing the STDM (Appendix C gives an overview of the spatio-temporal data model) using the object-oriented method proposed by Booch has elucidated several aspects concerning the time geography framework. First of all, the design decision to model events and states as object classes may not be appropriate in another application. Instead it might require events to be modelled as methods rather than object classes. This would certainly affect the design of the space-time paths within the model. But having decided to model events and states as object classes, the class diagrams in this chapter are valuable tools for analysis and design.

Second, the design of a space-time path has been defined at the instance level (object level). Different states, as well as events, have been defined as instances of classes having their respective relationships. However, the space-time path is only generated by connecting the instances of these object classes. This connection has been designed using an incremental modification mechanism. Unfortunately, Booch's method does not provide a diagram for modelling different scenarios at the instance level of object classes. However, the interaction diagrams have been used to illustrate how these scenarios are executed, as well as to visualise the process involved in 'making space-time paths' within the STDM.

Finally, the analysis and design of the STDM using an object-oriented method has raised issues in version management. Having a system with several instances of several object classes connected to space-time paths has definitely raised a question mark about versioning. What is the best approach for managing versions in order to avoid inconsistency in the representation of space-time paths within the STDM? In investigating the possibilities of designing a version management mechanism in the STDM, one of the main findings was the need to understand the meaning of 'change'. Change is responsible for the existence of states on a space-time path, and their interrelations with events. In other words, change carries out actions that create different instances in the space-time paths within the system. Therefore, change has to be well understood as an element of the version management approach of the STDM.

The next chapter describes the implementation aspects for incorporating the STDM into a GIS. The prototype implementation has been undertaken mainly as a 'proof-of-concept' for the ideas developed in the STDM.

CHAPTER SIX

Implementation of
the STDM

This chapter presents an example application to demonstrate the most important aspects of the STDM implementation. The four basic scenarios – public boundary entry, evolution tracking, update, and archiving – have been used to illustrate it. By understanding the individual scenarios it is possible to understand the whole application. The overall aim is to show how space-time paths can be created within a GIS. Some sample data sets of the East Cambridgeshire region have been used to illustrate the boundary-making process; they are provided from within Smallworld GIS. The examples are fictitious as the principal aim is to create a prototype implementation.

6.1   SMALLWORLD GIS

'How are applications developed in Smallworld GIS?' asks Easterfield (1993). And his reply: 'There are probably as many answers to this question as there are Smallworld users'. It is true that implementing a time geographic spatio-temporal model into an existing GIS is fraught with a whole assortment of problems essentially related to the GIS structure and its software configuration. Smallworld GIS was chosen for its static and dynamic configurations, which have been previously designed for the STDM. The static configuration is related to the object properties (attributes) and the relationships among objects. In Smallworld the definition of object properties and their relationships for a particular application are loaded into the VMDS (version-managed data store) using the Case tool. The only type of structure in the VMDS is the table, except for one block which forms the hierarchy of indices mapping different versions of the database.

On the other hand, the dynamic configuration is related to the operations on objects and properties. In Smallworld GIS the methods for a particular application are loaded into the VMDS using the exemplars file – a Magik source code file defining the method. Magik is the programming environment in Smallworld GIS; it consists of methods and classes which are independent of those existing in the VMDS. In this environment, system programming, application development, system integration and customisation are all written using the Magik programming language. Bearing in mind both static and dynamic configurations, the desirable object-oriented features for implementing the STDM in Smallworld GIS have been defined as follows:

- *Object identifiers* allow an object to be referenced via a unique internal generator number.
- *References attributes* are used to represent relationships between objects. Although they are analogous to pointers in a programming language or to foreign keys in a relational system, there are two important differences. References attributes cannot be corrupted, whereas pointers can be. And references attributes are not associable with a user-visible value, whereas foreign keys are.
- *Collection attributes* such as LIST, SET or ARRAY of values.
- *Derived attributes* are defined procedurally rather than stored explicitly. A procedure is specified to be executed when the value is retrieved or assigned.
- *Referential integrity* defines the correctness of references when an object is deleted or a relationship is changed. There are five levels of referential integrity (Cattel, 1991):
  1   No integrity checks.
  2   The system may delete objects automatically when they are no longer accessible by the user, e.g. garbage collection algorithms in the GemStone system.
  3   The system may require that objects are deleted explicitly when they are no longer used, but may detect invalid references automatically. An example is the IRIS system.
  4   The system allows explicit deletion and modification of objects and relationships, and may automatically maintain the correctness of relationships as seen through all objects. Two examples are the ONTOS and Probe systems.
  5   The system allows the database designer to specify customised referential integrity semantics for each object or relationship.
- *Aggregation relationships* group parts into a whole.
- *Inheritance relationships* exist between objects.
- *Methods* are associated with objects.

In investigating the support provided by Smallworld GIS for these features, an important distinction can be made between a Magik image environment and a data store view of VMDS. This is because some features are available only in a Magik image, and others only in a data store view, and very few are available in both image and view. Table 6.1 summarises the main findings about the object-oriented features available in Smallworld GIS. Although a Magik image is associated with a data store view within Smallworld GIS, the available object-oriented features are distinct between them. The Magik image holds more object-oriented capabilities

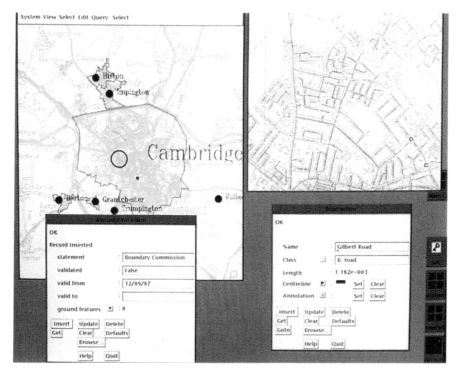

Figure 6.2   Execution of the public boundary entry scenario – stage 2

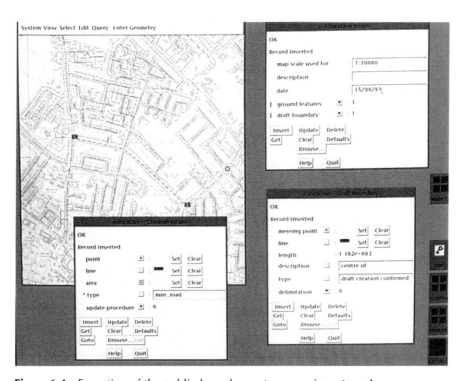

Figure 6.4   Execution of the public boundary entry scenario – stage 4

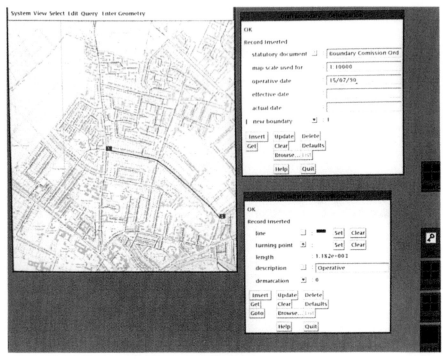

Figure 6.6 Execution of the evolution tracking scenario – stage 2

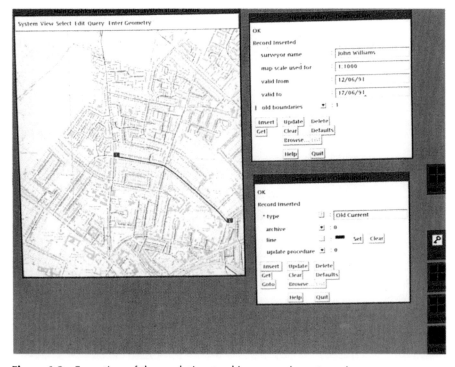

Figure 6.8 Execution of the evolution tracking scenario – stage 4

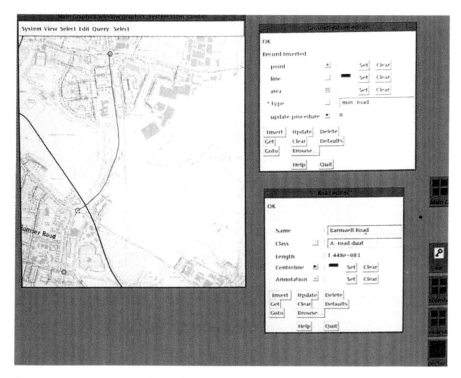

Figure 6.9 Execution of the update scenario – stage 1

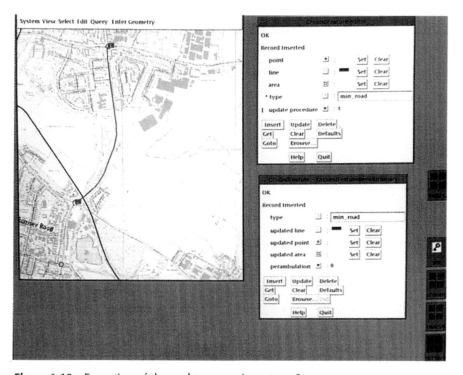

Figure 6.10 Execution of the update scenario – stage 2

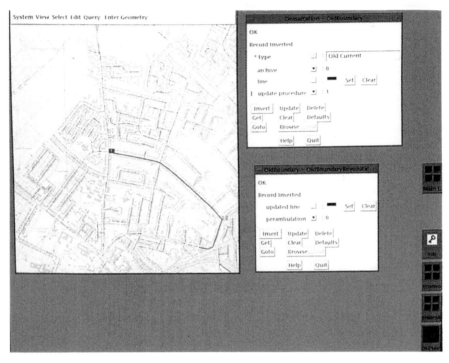

Figure 6.12 Execution of the update scenario – stage 4

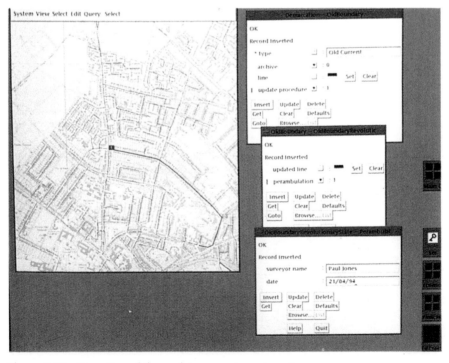

Figure 6.13 Execution of the update scenario – stage 5

Table 6.1   Smallworld GIS: object-oriented features.

Feature required by the STDM	Magik image	Data store view
Allow direct access to object identifiers	yes	yes
Allow the object to have meaningful object keys[a]	yes (object keys)	yes (primary keys)
Allow the use of references attributes	no	no
Allow collection attributes (list, set, array)	yes	no
Allow derived attributes	no	yes
Allow update of derived attributes	no	no
Allow referential integrity	yes (level 2)	yes (level 4)
Allow aggregation relationships	no	no
Allow inheritance	yes (multiple inheritance)	no
Allow polymorphism	yes	no
Allow methods associated with objects	yes	yes
Allow free methods	yes (procedures)	no

[a] Equivalent to primary keys in the relational model.

than a data store view of the VMDS. This indicates the existence of a proprietary database (VMDS) being managed by an object-oriented environment (Magik).

6.2   PUBLIC BOUNDARY ENTRY SCENARIO

The public boundary entry scenario represents the states and events that are concerned with the allocation events within the system. This means that, based on some assumptions, the user will select ground features to be a future public boundary. The final allocation decision will be taken by the user, who will assign a ground feature to be a draft boundary. The execution of this scenario has been divided into four stages, illustrated by diagrams. The first stage focuses on the creation circumstance of a space-time path of a public boundary. It represents the origin of a space-time path within Smallworld GIS.

Stage 1: creation

In our example the origin has been attached to the assumption event of the STDM. The user has made a statement perhaps based on a boundary commission demand which determines that a possible future public boundary will be created in a certain region. Figure 6.1 illustrates this situation in which the Assumption menu (bottom

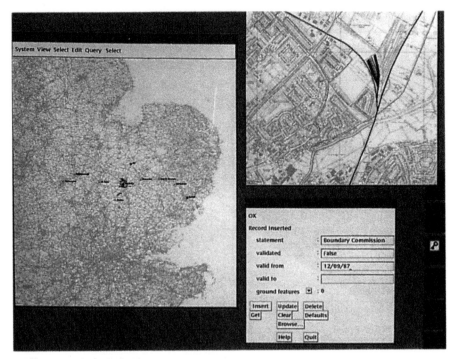

Figure 6.1   Execution of the public boundary entry scenario – stage 1

right) represents the assumption event involved in the creation of a space-time path. It shows the properties previously defined for the Assumption class of the STDM: statement, validated, and valid from/to. In this example the boundary commission demand for creating a public boundary has not been validated by an order or act. This demand has been valid since 12 September 1987.

The user can select any instance that belongs to the GroundFeature, DraftBoundary, NewBoundary or OldBoundary classes, or any instance that belongs to the Assumption, Allocation, Delimitation or Demarcation classes. The origin of a space-time path in the VMDS can never be modified. In this example the origin of a space-time path is created once the properties of the Assumption class have been inserted in the data store view of the VMDS. This is still true even though the space-time path does not exist at this stage. In other words, the relationship between the instance of the Assumption class and the instance of the GroundFeature class has not yet been generated in Smallworld GIS. Therefore the ground features element in the Assumption menu shows the number 0.

Stages 2, 3: selecting a ground feature

In stage 2 the user is searching for the ground feature designated by the boundary commission, e.g. Gilbert Road, which is deemed to belong to the proposed public

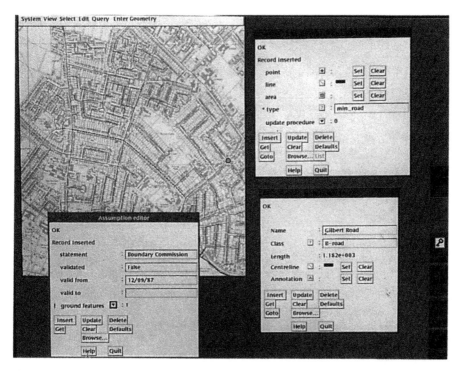

Figure 6.3   Execution of the public boundary entry scenario – stage 3

boundary. This is accomplished by querying where Gilbert Road is located by using the Road Editor menu provided by Smallworld GIS. In Figure 6.2 (see colour section) the Road Editor menu (bottom right) shows the properties that belong to Gilbert Road. The Application menu (top right) highlights where this road is located.

Once Gilbert Road has been identified on the screen, the user can then make the assumption that this road can be a public boundary as a result of the boundary commission demand. This is achieved by clicking on the arrow beside the ground features element in the Assumption menu (Figure 6.2, bottom left), which will activate the Assumption–GroundFeature menu (Figure 6.3, top right).

In fact, the activated Assumption–GroundFeature menu is the state of an instance in the GroundFeature class. It shows the properties of the Ground-Feature class: point, line, area and type. The insert operation is invoked when the user assigns the values of the properties of the GroundFeature class and inserts them into the VMDS by clicking on the Insert element of the Assumption–GroundFeature menu. After creating the state of an instance in the Ground-Feature class, the ground features element in the Assumption menu (Figure 6.3, bottom left) is updated to the number 1, telling the user that one state of a ground feature has been created for the specific assumption. In other words, there exists a space-time path between the instances of the Assumption class and the GroundFeature class. In terms of Smallworld GIS, there exists a *join* relationship between the records of the Assumption class and the Ground-Feature class.

This implies that the version significant attributes (point, line and area) of the GroundFeature class can only be updated in a non-destructive manner (see Table 5.5). An update in one of these attributes pertaining to the GroundFeature class will create a version that will be the new instance of the Ground-FeatureRevolutionaryState class. The update procedure element of the Assumption–GroundFeature menu shows the number 0, which informs the user that no update procedure has been carried out. Section 6.4 considers update procedures in more detail.

Stage 4: allocation

In stage 4 the allocation decision is taken by the user, so the ground feature is selected to be a draft boundary. In terms of the STDM, the GroundFeature, Allocation and DraftBoundary classes are affected by this allocation decision. The execution of this scenario is accomplished by activating the Allocation menu (Figure 6.4, top right – see colour section) which represents the Allocation class within the system. The user can provide information about the map scale used for the allocation, a textual description of the allocation and the date on which the allocation took place (Figure 6.4, colour section).

The Allocation–GroundFeature menu (Figure 6.4, bottom left – see colour section) shows that the user has selected Gilbert Road to be a draft boundary. The Application menu (Figure 6.4, top left) highlights where this road is located. The user can then assign this ground feature to be a draft boundary. This is achieved by clicking on the draft boundary element of the Allocation menu, which will activate the Allocation–DraftBoundary menu (Figure 6.4, bottom right). The activated menu represents the instance of the DraftBoundary class. The user has inserted the attribute values into the VMDS by clicking on the Insert element of the Allocation–DraftBoundary menu. In this case all attributes of the DraftBoundary class have been defined as non-version significant attributes in the STDM (see Table 5.5). They can be updated without creating a new version.

The available elements of each of the menus displayed at this stage (Figure 6.4) are as follows:

- The ground features element shows the number 1, indicating that Gilbert Road has been selected to be a draft boundary.

- The draft boundary element shows the number 1, indicating that Gilbert Road has been allocated to be a draft boundary.

- The update procedure element shows the number 0, indicating that no update procedures have been carried out so far.

- The delimitation element shows the number 0, indicating that the delimitation process has not yet occurred at this stage.

The STDM design, which governs the behaviour of the states on a space-time path, does impose some constraints on this scenario, e.g. coupling constraints and capacity constraints. For example, a draft boundary state will never be created if its corresponding ground feature state does not previously exist in the VMDS of Smallworld GIS. In the same way, an allocation cannot take place without the previous existence of a ground feature state. This has been achieved by defining update methods for the invariant and version significant attributes of the relevant classes.

6.3   EVOLUTION TRACKING SCENARIO

Both delimitation and demarcation events are described in the evolution tracking scenario. In the delimitation event, a draft boundary is confirmed by an act or order, so the draft boundary state becomes a new boundary state. For the demarcation event, surveyors ratify the position of the new boundary on the ground, then the state is changed from new boundary to old boundary. The execution of this scenario has been divided into four stages. The DraftBoundary, Delimitation, NewBoundary, Demarcation and OldBoundary classes of the STDM are used to illustrate our example. The focus is on demonstrating the independent incremental modification mechanism developed for the space-time path of the STDM. Since the inheritance mechanism is not supported by the data store view of Smallworld GIS, the independent incremental mechanism has been implemented through the *join* relationships between the classes. The aim is to illustrate how a user could handle space-time paths in a GIS.

Stages 1, 2: delimitation

Stages 1 and 2 focus on the delimitation event for the existence circumstance of a space-time path. Continuing with the Gilbert Road example, the draft boundary state has been selected by the user, as shown from the Allocation–DraftBoundary menu (Figure 6.5, top right); the draft boundary is highlighted in the Application menu (Figure 6.5, left).

The delimitation event is activated within the system by clicking on the delimitation element of the Allocation–DraftBoundary menu. The Draft-Boundary–Delimitation menu appears (bottom right) containing the relevant attributes that belong to the Delimitation class of the STDM. These attributes include the statutory document and the map scale used for the delimitation, as well as the operative, effective and actual dates related to the delimitation event.

Once the attribute values of the Delimitation class have been inserted into the VMDS, the delimitation element of the Allocation–DraftBoundary menu shows the number 1. This number indicates as many delimitations as have been inserted by the user into the VMDS. On the other hand, the new boundary element of the DraftBoundary–Delimitation menu displays the number 0 since the new boundary state has not yet been created in the VMDS.

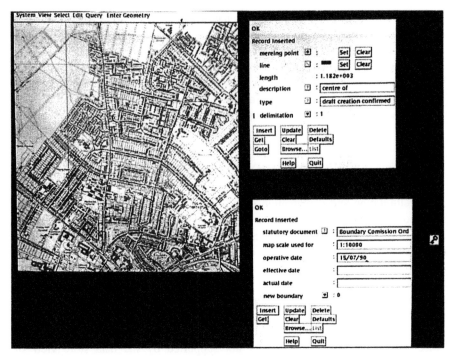

Figure 6.5   Execution of the evolution tracking scenario – stage 1

Once the user has decided to create the new boundary state in the system, they have to click on the new boundary element of the DraftBoundary–Delimitation menu in order to activate the Delimitation–NewBoundary menu (Figure 6.6, bottom right – see colour section). This activated menu represents a new instance of the NewBoundary class.

A state is created when the user inserts the values of the attributes of the NewBoundary class by clicking on the Insert element of the Delimitation–NewBoundary menu. As a result, the new boundary element of the Draft-Boundary–Delimitation menu (Figure 6.6, top right) is automatically updated to the number 1. The user is then aware of the existence of a new boundary state in the VMDS of Smallworld GIS. In this case all attributes belonging to this new boundary state have been defined as non-significant attributes within the STDM. Their updates do not create versions in the system.

Stages 3, 4: demarcation

The description of stages 3 and 4 has been devised to avoid mentioning the conceptual elements behind the application. This has been important in demonstrating how simple and repetitive are the stages in building up space-time paths within Smallworld GIS. In general, the user has only to follow the successor-in-time creation of different

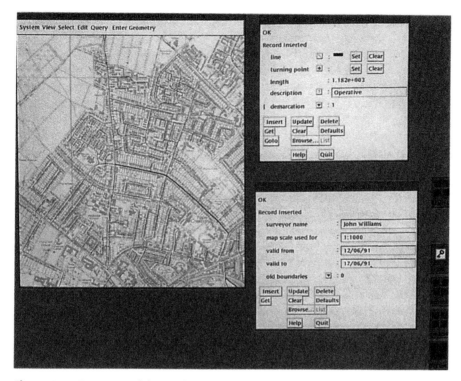

Figure 6.7   Execution of the evolution tracking scenario – stage 3

states within the system. A newer state can only be created if its corresponding older state in the space-time path already exists.

Stage 3 denotes when the demarcation event takes place within the system. This involves the creation of an old boundary state from its corresponding new boundary state. The user clicks on the demarcation element of the Delimitation–New-Boundary menu, activating the display of the NewBoundary–Demarcation menu (Figure 6.7, bottom right). This menu displays the relevant information about the demarcation event, including the properties defined for the Demarcation class: the name of the surveyor in charge of carrying out the demarcation, the map scale used, and the time taken by the surveyor to demarcate the boundary on the ground.

At this stage, the old boundary state has not yet been created in the system. The old boundaries element in the NewBoundary–Demarcation menu shows the number 0. The user activates the Demarcation–OldBoundary menu (Figure 6.8, bottom right – see colour section) by clicking the old boundaries element in the NewBoundary–Demarcation menu. Once the attributes related to an old boundary state have been inserted into the VMDS by the user, the old boundaries element in the NewBoundary–Demarcation menu (Figure 6.8, top right) shows the number 1.

The Demarcation–OldBoundary menu shows two important elements, the archive element and the update procedure element. The archive element, if activated, will archive the selected old boundary. This is discussed in more detail in Section 6.5. The `update procedure` element indicates that no update procedure has been carried out over the selected old boundary state. This is discussed in more detail in the following section. Both `NewBoundary` and `OldBoundary` classes have different display scales as designed in the STDM. They have been implemented using the Magik programming environment of Smallworld GIS.

6.4   UPDATE SCENARIO

The update scenario handles the update procedures due to natural changes and new demarcation descriptions occurring over public boundaries. Three update procedures have been defined as creation of a new object, creation of a new object from an existing object, and relocation of an existing object (Chapter 5, Section 3.3). The aim here is to illustrate the main aspects involved in creating the versions due to these update procedures, by using the overlapping incremental modification of the STDM. The first update procedure has already been considered, so this section concentrates on the second update procedure (stages 1 to 3) and the third update procedure (stages 4 and 5).

Stages 1, 2, 3: creation from an existing object

Suppose the user has selected as ground feature the Barnwell Road object, as shown in the Road menu (Figure 6.9, bottom right – see colour section) and highlighted in the Application menu. The GroundFeature menu (Figure 6.9, top right) shows the instance of the `GroundFeature` class that represents Barnwell Road. In fact, this represents the instance of this class which is about to be updated. The update procedure creates a new version in the `GroundFeatureRevolutionaryState` class using the previous version in the `GroundFeature` class.

The user can perform the update procedure by clicking on the `update procedure` element of the GroundFeature menu. This will activate the Ground-Feature–GroundFeatureRevolutionaryState menu (Figure 6.10, bottom right – see colour section). This activated menu represents the version (i.e. the instance of the `GroundFeatureRevolutionaryState` class) which is created in such a way that the invariant attribute (i.e. type) of the `GroundFeature` class is inherited by the new instance of the `GroundFeatureRevolutionaryState` class (Figure 6.10, colour section). In this example the inherited attribute is min_road type.

The version significant attributes (point, line and area) belonging to the `GroundFeature` class are not inherited. In order to differentiate, the `GroundFeatureRevolutionaryState` class presents updated point, updated line and updated area as attributes. All the attributes of the `GroundFeatureRevolutionaryState` class are now considered as invariant attributes. They cannot be modified or deleted by any user at any time.

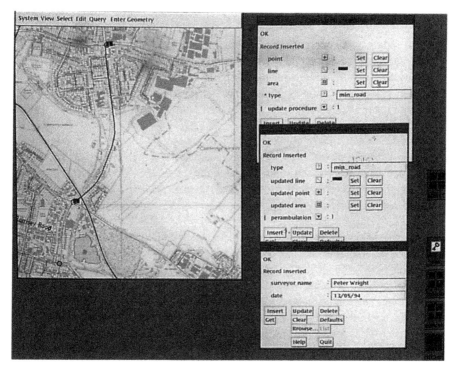

Figure 6.11   Execution of the update scenario – stage 3

The user creates a new version of the ground feature object, in this case Barnwell Road, by digitising its new position and inserting it as the updated line attribute of the GroundFeatureRevolutionaryState class. Once the user has inserted the new updated line for the GroundFeatureRevolutionaryState class, the update procedure element of the GroundFeature menu (Figure 6.10, top right) is changed to 1, indicating that one update has occurred to this particular ground feature. The perambulation element of the GroundFeature–GroundFeatureRevolutionaryState menu indicates the number 0, which signifies that the user has updated the ground feature on the system, but this has not yet been confirmed on the landscape by the surveyors.

The perambulation event occurs after the surveyors have confirmed that the change in position of the ground feature has taken place on the landscape. The user can then insert this information into the system by clicking on the perambulation element of the GroundFeature–GroundFeatureRevolutionaryState menu. This will activate the Perambulation menu (Figure 6.11, centre right) which contains the attributes concerning the perambulation event, e.g. the name of the surveyor and the date when the perambulation occurred. At this stage, the perambulation element of the GroundFeatureRevolutionaryState menu indicates that one perambulation event has occurred for the selected ground feature revolutionary state.

Stages 4, 5: relocation of an existing object

The last update procedure in the STDM involves the relocation of an existing object. The same boundary used for illustrating the evolution tracking scenario has been chosen to illustrate this update procedure in the update scenario. Figure 6.11 shows the relocation of this boundary. This is illustrated by the Demarcation–Old-Boundary menu (Figure 6.12, top right – see colour section), in which the state of this boundary is displayed. The update procedure has been carried out by activating the OldBoundary–OldBoundaryRevolutionaryState menu (Figure 6.12, bottom right) using the update procedure element of the Demarcation–OldBoundary menu.

The user has inserted the new updated line into the VMDS, so the update procedure indicates the number 1. The OldBoundary–OldBoundaryRevolutionary-State menu represents the invariant attributes of the instance of the Ground-FeatureRevolutionaryState class.

Once the old boundary is relocated and its new updated line is inserted into the system, the instance of the old boundary revolutionary state is created. This state presents only invariant attributes, and it cannot be updated again. Once the confirmation of the public boundary's new position in the landscape is obtained, the user can insert the relevant information about the perambulation event by clicking on the perambulation element of the OldBoundary–OldBoundaryRevolutionaryState menu. The OldBoundaryRevolutionaryState–Perambulation menu (Figure 6.13, bottom right – see colour section) is activated, and the user is then able to capture the attribute values of the perambulation event within the system.

6.5   ARCHIVING SCENARIO

The archiving scenario represents the archiving of old boundaries that are no longer effective. It involves the creation of an obsolete state during the lifespan of a public boundary. This is achieved by clicking on the archive element of the OldBoundary menu (see Figure 6.8). Smallworld GIS allow users to store obsolete boundary states on a mass storage device such as a CD-ROM, maintaining the other states on a conventional hard disk.

6.6   HISTORICAL VIEWS

Historical views have been designed in the spatio-temporal data model in order to visualise the incremental mechanism of space-time paths. They provide direct access to historical data concerned with a particular public boundary. This has been achieved by creating HistoricalView classes in Smallworld GIS. The independent incremental mechanism of the STDM is illustrated in Figure 6.14. Two historical views are displayed as the Delimitation Historical View menu and the Demarcation Historical View menu. Both menus represent the union of all proper-

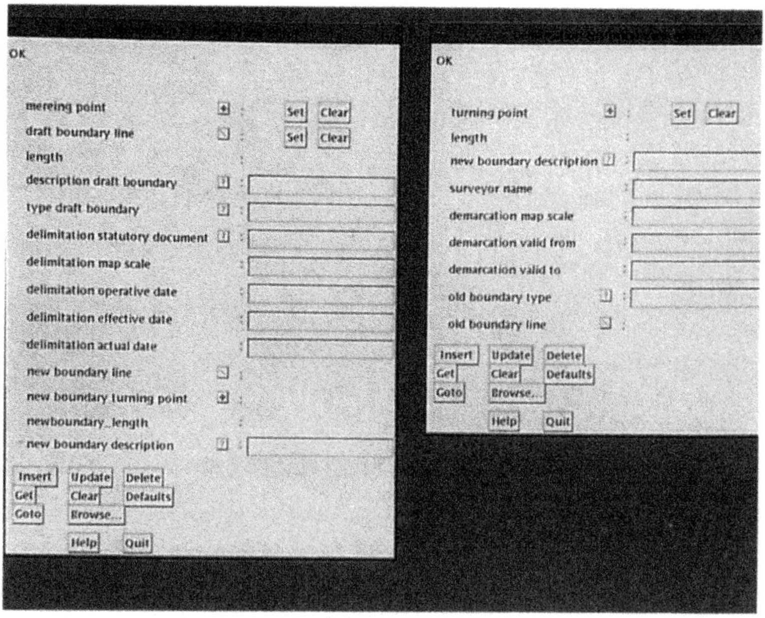

Figure 6.14 Historical views for the evolution tracking scenario

ties that belong to the parent, modifier, and resultant classes in the evolution tracking scenario. The historical view provides a snapshot of all properties of a specific public boundary which are selected by the user through a query statement. In Figure 6.14 the delimitation historical view provides the selection of all properties that belong to `DraftBoundary`, `Delimitation` and `NewBoundary` classes.

The overlapping incremental mechanism of space-time paths has been illustrated by creating a historical view termed update historical view (Figure 6.15). In this case the Update Historical View menu illustrates the selection of properties from the `GroundFeature`, `GroundFeatureRevolutionaryState`, `Perambulation`, `OldBoundary` and `OldBoundaryRevolutionaryState` classes.

In fact, the user can create as many historical views as they need for their application. And they can specify the classes that should belong to each historical view. However, further research work is needed to address the historical views effectively. Geographic visualisation of historical views may provide a tool to assist the user in displaying the historical views in graphic, symbolic and ideally dynamic forms, such as animated graphics, graphs or diagrams, or even a combination of them.

Geographic visualisation may generate a great deal of graphical information (McCormick, Defanti and Brown, 1987) so users need to have a natural acuity for recognising and interpreting visual patterns (Fedra, 1992; Buttenfield, 1993), and an intuitive understanding of large amounts of data, processes and interdependencies

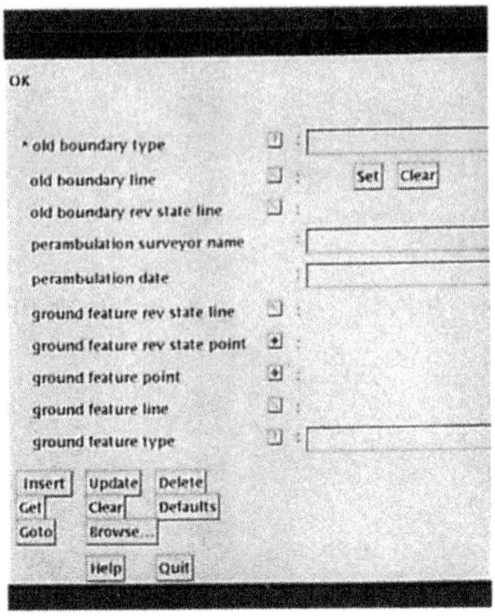

Figure 6.15   The update historical view

between historical views. The role of geographic visualisation is to provide the means for a better understanding of the STDM and the information stored in the database.

6.7   CONCLUSIONS

This chapter has illustrated some examples of how space-time paths can be created by a user. The states and events designed in the STDM are visualised as edit menus within the system. This allows a better understanding of how space-time paths can be implemented in Smallworld GIS. The prototype implementation has been undertaken mainly as a 'proof-of-concept' of the ideas developed in the STDM. Several drawbacks have been identified in implementing this prototype into Smallworld:

- The data store view of the VMDS does not allow references attributes or inheritance relationships between objects (see Table 6.1). Foreign keys have had to be used to implement the *join* relationships.

- Because the Magik image and the data store view support different levels of referential integrity (see Table 6.1), the prototype implementation has shown a discrepancy between these levels when they are executed for checking the correctness of the relationships between states and events on space-time paths. The garbage collection algorithms in the Magik image are more resilient as a referential integrity approach. They detect a greater variety of errors than the checking support in a data store view.

- Although historical views have been implemented in the system, a geographic visualisation tool has proved essential for displaying the spatio-temporal data that belong to a historical view. Further development work would be required to handle historical views fully within the system.

The prototype implementation has also indicated possible improvements in the STDM:

- The STDM captures the essential semantics (space-time paths, events and states) of an application, and through this model the user sees their problem domain composed of objects and the classes to which they belong. Classes and their relationships may be further refined, and some classes may be added or deleted. The integration of time geography and object orientation within a GIS has proven to be a dynamic modelling activity, in which the schema is constantly being upgraded.

- The constraints in the STDM are simple. They are methods attached to the instances of each class. This could be improved by associating knowledge-based rules with events and states of the STDM. For example, 'when' event 'if' condition 'then' action.

CHAPTER SEVEN

Emerging technologies

This chapter considers four emerging research topics related to geographical information sciences. They are some of the latest developments in the research fields of databases, geographic visualisation and distributed systems. Each section begins with a brief description of the research area then examines its impact on designing spatio-temporal data models in GIS.

7.1 SPATIO-TEMPORAL OBJECTS IN DATABASE SYSTEMS

Despite having interrelated aims, research in temporal and spatial database models has predominantly developed independently. Research in temporal database design (Al-Taha, Snodgrass and Soo, 1994; Snodgrass, Al-Taha and Soo, 1993; Tansel *et al.*, 1993; Soo, 1991) has addressed issues related to version management aspects (time stamping, concurrency control, update processing) as well as enhancements required for the logical components (schema evolution, temporal query language syntax) and the physical structure (storage structure, access methods, query optimisation, query language features). Spatial database research has focused on the application-specific semantics of handling change in spatial data models. In particular the issues related to (1) the support of changes in topological relationships among entities; (2) the dynamic representation and visualisation of versions; and (3) the support for spatial data types and spatial queries (Armenhakis, 1993; Egenhofer and Al-Taha, 1992; Svensson and Huang, 1991).

Only recently has there been an increase in research efforts for combining spatial and temporal database developments. One of the most significant contributions is the innovative approach proposed by Erwig, Schneider and Güting (Erwig, Schneider and Güting, 1997; Erwig *et al.*, 1997) with the Chorochronos project. This approach offers a broader spectrum of integration by embedding the combination of temporal and spatial objects into *spatio-temporal objects* in databases. Spatio-temporal objects describe the temporal behaviour of 'moving' points and regions within the model. They are regarded as three-dimensional (2D space + time) or higher-dimensional entities whose structure and behaviour is captured by modelling them as abstract data types (ADTs). These abstract types can be implemented as attribute data types of an object into an object-oriented data model. The approach proposes the role of spatio-temporal data types as being fundamental to the role played by spatial data types in spatial databases.

S*patio-temporal objects* in a spatio-temporal data model capture the synergy between space and time of a *space-time path* in Time Geography. In terms of STDM, spatio-temporal data types offer the advantage of defining object classes with properties that are simultaneously defined in space and time (2D space, time). An event in space and time can be defined as (point, point). Otherwise an event having a certain period of time could be defined as (point, interval). Likewise, a set of events can be defined as a *sequence of (point, point)* as well as a *sequence of (point, interval)*. A region event in space and time can be defined as (region, point) as well as (region, interval). As far as spatio-temporal data modelling is concerned, this approach provides an innovative way of dealing with space-time paths in the STDM. Further exploration is required into the modelling strengths of spatio-temporal objects for creating space-time paths. In particular, spatio-temporal data types are more versatile and offer much more control over temporal behaviour of objects. Future work should consider the study of spatio-temporal data types with spatio-temporal analytical operations.

7.2   KNOWLEDGE DISCOVERY IN DATABASES

A major challenge for researchers investigating the next generation of geographic information systems is to develop methods for finding useful information such as patterns, trends and correlation from massive spatio-temporal data stored in the database. The development of knowledge discovery in databases (KDD methods) has recently emerged from research carried out in the areas of database systems, artificial intelligence, machine learning, statistics and geographic visualisation (Glymour *et al.*, 1997; Fayyad *et al.*, 1996). They are focused on developing a complex interactive and iterative process of identifying patterns that are novel and useful for a knowledge domain. One of the main steps involved in a KDD process is data mining. Three subcomponents can be distinguished for the data mining step: (a) choosing the data mining task, e.g. classification, clustering, association and pattern-based similarity search; (b) choosing the data mining algorithm to perform the mining task; and (c) applying the algorithm to a target data set.

One particular mining task is relevant to designing space-time paths for a knowledge domain – extracting classes from target data. This classification process involves the search for common attributes among a set of objects, and then the arrangement of these objects into classes according to a partitioning criterion, model or rule. GIS researchers are particularly interested in unsupervised data mining algorithms designed to uncover classes in spatio-temporal data, working under the assumptions that the class labels are a priori unknown. They are of fundamental importance to the analysis and design of spatio-temporal data models. They provide a computable representation of how the spatio-temporal data can be distributed into classes in a way that was not previously conceived for the database. In other cases they can be used to predict new objects of classes from outside the database and they can be used to develop a predictive spatio-temporal data model.

Class extraction implies that new and previously unknown patterns will be detected in the spatio-temporal data. In terms of future research in spatio-temporal data modelling, the focus is on the efficiency of implementing data mining algorithms in GIS. This would require the development of spatio-temporal data mining algorithms for KDD, processing techniques for spatio-temporal queries, and geographic visualisation of the KDD process. Besides, the reliability assessment of KDD methods is of fundamental importance to data miners. The main issue is uncovering under what conditions a search algorithm provides correct classification for designing a spatio-temporal data model. Even the best KDD methods of search and statistical assessment leave the data miner with a range of uncertainties about the correct classification or the correct prediction. This opens a vast research area into spatio-temporal data modelling for quantifying these uncertainties and their association with the creation of classes, and subsequently with the creation of space-time paths within the STDM.

7.3 GEOGRAPHIC VISUALISATION

A fully developed KDD process is not anticipated in the foreseeable future. In fact, Brachman and Anand (1996) point out that a realistic implementation of KDD is only possible as a human-centred process that is characterised by human visual thinking, computer data animations and human–computer interaction. Geographic visualisation (GVis) research plays an important role in developing such a human-centred process in KDD. Specifically the GVis research has been focused on (a) understanding the iterative nature of human interaction with visual displays of spatio-temporal data (Dykes, 1997; MacEachren and Taylor, 1998) and (b) geographic visualisation methods designed to integrate graphical interactive exploratory analysis with GIS (Cook et al., 1997).

Geographic visualisation involves much more than just enabling users to 'see' spatio-temporal data. Users must be able to visualise the data and focus on what is relevant. Users also need to communicate and share information in collaborative settings and act directly to perform exploratory analysis tasks based on this information. This involves the support of information search, analysis, communication and systems control operations within a single interactive user interface.

The potential integration of GVis, GIS and KDD into a comprehensive system that provides interactive visual displays, spatio-temporal operations and data mining capabilities is fundamental to the development of spatio-temporal data models, but unfortunately it has still to be realised in theory let alone in practice. One of the major contributions towards this integration is found in the research work developed in the Apoala Project (MacEachren et al., 1999; Qian et al., 1997;

Wachowicz *et al.*, 1998). The approach focuses on the integration of GVis, GIS and data mining operations in the context of spatio-temporal data on the earth's environment.

7.4   UBIQUITOUS COMPUTING

The term 'ubiquitous computing' signifies the ability of multi-platform systems to interact with each other through the interchange of data and functions. These systems are often known as open systems to indicate their interoperability in developing a common language for building data and communication models among cooperative applications. Such a language model provides a way unambiguously to specify system requirements (e.g. desirable features, executable functions and extensibility). Communication models are often modelled using an object-oriented approach. An object encapsulates state and provides a well-defined interface to the rest of the system. Objects interact by invoking operations on these interfaces.

The Open GIS consortium (OGIS) is an independent body with members from academia and industry that has as its main objective the standardisation of system specification, modelling and implementation of interoperable GIS. The OGIS model defines spatial and temporal interfaces, specifies geodata definitions of spatial, temporal and spatio-temporal domains, feature, coverage and attributes. The OGIS model also defines interoperable services that are designed to access heterogeneous spatio-temporal databases. Services will provide the mechanisms to locate, retrieve and update items of data, regardless of the structures involved. The user will be able to access data without being concerned with details of data formats and format conversion.

The overall architecture of OGIS is based on an object-oriented implementation of existing distributed computing systems. Everything in OGIS is an object. Every component of the OGIS model (e.g. data format, service, communication and application code) can be manipulated as an object. The result is interchangeable software parts that can be implemented to fit specific user needs. Application developers will have the flexibility to select a set of software parts without the need fully to implement a GIS. The choice will be based on selecting the implementation that is best suited to the task required by the user. Consequently, the development of spatio-temporal data models based on object-oriented analysis and design methods is fundamental to creating an open GIS. The STDM provides the user with a coherent computational model that encapsulates object classes, and spatio-temporal data may be established and controlled by interaction of the objects. Additional work is required to support the concept of services within the model. In particular, the semantics and mechanisms for sharing data and operations need to be understood. For example, communication services for sharing multi-scale and multi-dimensional data in a distributed system environment, and interaction services for sharing GIS, GVis and data mining operations as required by a user-defined task.

Notation for Booch's method

Appendix A is reprinted, with permission, from Booch, G., 1994, *Object-Oriented Analysis and Design with Applications*, 2nd edn, Santa Clara CA: Benjamin/Cummings. The maps in Appendix B are courtesy of L.J. Rackham.

CLASS DIAGRAM NOTATION

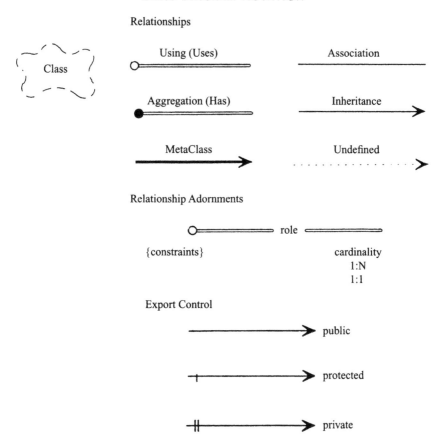

Relationships

Relationship Adornments

{constraints}   role

cardinality
1:N
1:1

Export Control

public

protected

private

PROCESS DIAGRAM NOTATION

| Processor | Device | Connection |

Depicting public boundaries on OS basic scales

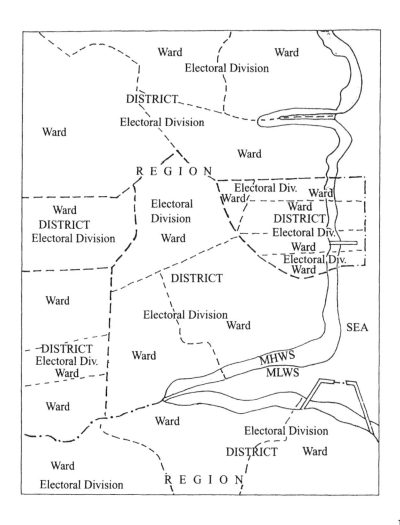

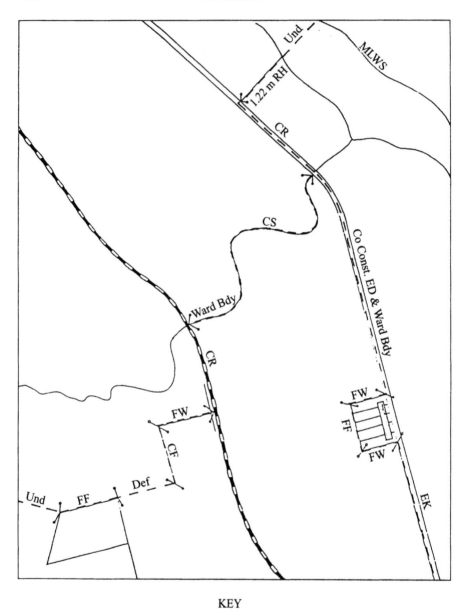

KEY

		CR = centre of road/rly	FW = face of wall

- - -   public boundaries   CR = centre of road/rly   FW = face of wall

- -|-   change of mereing   CF = centre of fence   RH = root of hedge

                                  CS = centre of stream   Def = defaced

MLWS   Mean Low Water   EK = edge of kerb   Und = undefined

            Spring Tides   FF = face of fence

Overview of the spatial data model

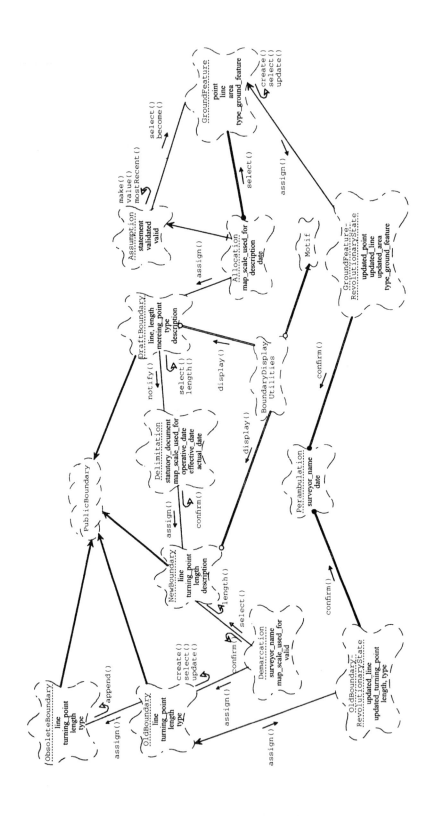

References

ACKOFF, R.L., 1981, *Creating the Corporate Future*, New York: John Wiley.

AHMED, R. and NAVATHE, S.B., 1991, Version management of composite objects in CAD databases, *Proceedings of the ACM SIGMOD International Conference on Management of Data*, pp. 218–27.

ALLEN, J.F., 1983, Maintaining knowledge about temporal intervals. *Communications of the ACM*, **26**, 832–43.

AL-TAHA, K.K. and BARRERA, R., 1990, Temporal data and GIS: an overview. *Proceedings of the GIS/LIS'90 Conference*, Vol. 1, pp. 244–54.

AL-TAHA, K.K., SNODGRASS, R.T. and SOO, M.D., 1994, Bibliography on spatio-temporal databases. *International Journal of Geographical Information Systems*, **8**(1), 95–103.

ANDERSON, T.L., 1982, Modeling time at the conceptual level, in SCHERERMANN, P. (ed.) *Improving Database Usability and Responsiveness*, Jerusalem: Academic Press, pp. 273–97.

ARMENHAKIS, C., 1993, Map animation and hypermedia: tools for understanding changes in spatio-temporal data, *Proceedings of the Canadian Conference on GIS*, pp. 859–69.

ARMSTRONG, M.P., 1988, Temporality in spatial databases, *Proceedings of the GIS/LIS'88 Conference*, Vol. 2, pp. 880–90.

BANERJEE, J., KIM, W., KIM, H.J. and KORTH, H.F., 1987, Semantics and implementation of schema evolution in object-oriented database systems, *Proceedings of the ACM SIGMOD Conference*, Vol. 5, no. 1, pp. 3–26.

BARRERA, R., FRANK, A.U. and AL-TAHA, K., 1991, Temporal relations in geographic information systems: a workshop at the University of Maine. *SIGMOD Record*, **20**(3), 85–91.

BEN-ZVI, J., 1982, The time relational model, PhD dissertation (unpublished), University of California, Los Angeles (UCLA).

BIAN, L., 1997, Modeling mobile objects in three-dimensional aquatic systems using object-oriented design, *Proceedings of GIS/LIS'97*, pp. 316–21.

BLOM, T. and LÖYTÖNEN, M., 1993, Research launch system to monitor epidemics in Finland. *GIS Europe*, **2**(5), 27–9.

BOEHM, B., 1986, A spiral model of software development and enhancement. *Software Engineering Notes*, **11**(4), 22–34.

BONCZEK, R.H., HOLSAPPLE, C.W. and WHINSTON, A.B., 1981, *Foundations in Decision Support Systems*, New York: Academic Press.

BOOCH, G., 1986, Object-oriented development. *IEEE Transactions on Software Engineering*, **12**(2), 211–21.

BOOCH, G., 1991, *Object-Oriented Design with Applications*, Redwood City CA: Benjamin/Cummings.

BOOCH, G., 1994, *Object-Oriented Analysis and Design with Applications*, 2nd edn, Santa Clara CA: Benjamin/Cummings.

BOOTH, J.R.S., 1980, *Public Boundaries and Ordnance Survey 1840–1980*. Southampton: R.A.G. Powell, Ordnance Survey.

BRACHMAN, R.J. and ANAND, T., 1996, The process of knowledge discovery in databases, in FAYYAD, U., PIATESTKY-SHAPIRO, G., SMYTH, P. and UTHURUSAMY, R. (eds) *Advances in Knowledge Discovery and Data Mining*, Menlo Park CA: AAI Press/MIT Press, pp. 37–57.

BURROUGH, P.A., 1986, *Principles of Geographical Information Systems for Land Resources Assessment*, Oxford: Clarendon Press.

BURROUGH, P.A. and FRANK, A.U., 1995, Concepts and paradigms in spatial information: Are current geographical information systems truly generic? *International Journal of Geographical Information Systems*, **9**(2), 101–16.

BUTTENFIELD, B.P., 1993, Scientific visualisation for environmental modeling: interactive and proactive graphics, *Proceedings of the Second Conference on Integrating GIS and Environmental Modelling*.

CARLSTEIN, T., PARKES, D. and THRIFT, N. (eds) 1978, *Timing Space and Spacing Time*, Lund: Royal University of Lund.

CATTELL, R.G.G., 1991, *Object Data Management: Object-Oriented and Extended Relational Database Systems*, Reading MA: Addison-Wesley.

CHANCE, A., NEWELL, R.G. and THERIAULT, D.G., 1990, An object-oriented GIS – issues and solutions, *Proceedings of the EGIS'90 Conference*, Vol. 1, pp. 179–88.

CHEN, P., 1976, The entity–relationship model: toward a unified view of data. *ACM Transactions on Database Systems*, **1**(1), 9–36.

CLARK, A., 1959, *Three Centuries and the Island*, Toronto: University of Toronto Press.

CLARK, A., 1962, The sheep/swine ratio as a guide to a century's change in the livestock geography of Nova Scotia. *Economic Geography*, **38**, 38–55.

CLIFF, A. and ORD, J., 1981, *Spatial Processes: Models and Applications*, London: Pion Press.

CLIFFORD, J. and ARIAV, G., 1986, Temporal data management: models and systems, in ARIAV, G. and CLIFFORD, J. (eds) *New Directions for Databases Systems*, Englewood Cliffs NJ: Ablex Publishing, pp. 168–85.

CLIFFORD, J. and WARREN, D.S., 1983, Formal semantics for time in databases. *ACM Transactions on Database Systems*, **8**(2), 214–54.

COOK, D., SYMANZIK, J., MAJURE, J.J. and CREPIER, N., 1997, Dynamic graphics in a GIS: more examples using linked software. *Computers & Geosciences*, **23**(4), 371–86.

COOMBES, M., OPENSHAW, S., WONG, C. and RAYBOULD, S. (1993), GIS in Community Boundary Definition, *Mapping Awareness and GIS*, **7**, 41–4.

COPELAND, G. and MAIER, D., 1984, Making Smalltalk a database system, *Proceedings of the ACM SIGMOD Conference*, pp. 316–25.

DADAM, P., LUM, V. and WERNER, H.D., 1984, Integration of time versions into a relational database system, *Proceedings of the Conference on Very Large Databases*, pp. 509–22.

DAHL, O.J., MYRHAUG, B. and NYGAARD, K., 1968, *Simula 67 Common Base Language*, Oslo: Norwegian Computing Centre.

DAHL, O.J. and NYGAARD, K., 1966, SIMULA – an Algol-based simulation language. *Communications of the ACM*, **9**, 671–8.

DEAN, T.L. and MCDERMOTT, D.V., 1987, Temporal data base management. *Artificial Intelligence*, **32**, 1–55.

DUTTON, G., 1987, *Proceedings of the First International Study on Topological Data Structures for Geographical Information Systems*, Reading MA: Addison-Wesley; *Harvard Papers on GIS*, Cambridge MA: Harvard University Press.

DYKES, J., 1997, Exploring spatial data representations with dynamic graphics. *Computers & Geosciences*, **23**(4), 475–82.

EASTERFIELD, M., 1993, Personal communication.

EFFENBERG, W.W., 1992, Time in spatial information systems, *First Regional Conference on GIS Research in Victoria and Tasmania*, Ballarat, Victoria.

EGENHOFER, M.J. and AL-TAHA, K.K., 1992, Reasoning about gradual changes of topological relationships, in FRANK, A.U., CAMPARI, I. and FORMENTINI, U. (eds) *Theories and Methods of Spatio-Temporal Reasoning in Geographic Space*, London: Springer-Verlag, pp. 196–219.

ERWIG, M., SCHNEIDER, M. and GÜTING, R.H., 1997, Temporal and spatio-temporal data models and their expressive power. *Informatik Berichte*, **225**(12), Fern Universität.

ERWIG, M., SCHNEIDER, M., GÜTING, R.H. and VAZIRGIANNIS, M., 1997, Spatio-temporal data types: an approach to modelling and querying moving objects in databases. *Informatik Berichte*, **224**(12), Fern Universität.

FAYYAD, U., PIATESTKY-SHAPIRO, G., SMYTH, P. and UTHURUSAMY, R. (eds) 1996, *Advances in Knowledge Discovery and Data Mining*, Menlo Park CA: AAI Press/MIT Press.

FEDRA, K., 1992, *Interactive Environmental Software: Integration, Simulation, and Visualisation*. IIASA RR-92-10, Vienna: International Institute for Applied Systems Analysis.

FISHER, P., 1997, Concepts and paradigms of spatial data, in CRAGLIA, M. and COUCLELIS, H. (eds) *Geographic Information Research: Bridging the Atlantic*, London: Taylor & Francis, pp. 297–307.

FRANK, A.U., 1994, Qualitative temporal reasoning in GIS – ordered time scales, *Proceedings of the SDH'94 Conference*, Vol. 1, pp. 410–30.

GARDELS, K., 1992, SEQUOIA 2000: new geographic information management technologies for global change research, *Proceedings of the EGIS'92 Conference*, Vol. 2, pp. 922–9.

GATRELL., A., 1983, *Distance and Space: A Geographical Perspective*, Oxford: Clarendon Press.

GLYMOUR, C., MADIGAN, D., PREGIBON, D. and SMYTH, P., 1997, Statistical themes and lessons for data mining. *Data Mining and Knowledge Discovery*, **1**, 11–28.

GOLLEDGE, R.G. and STIMSON, R.J., 1997, *Spatial Behaviour: A Geographic Perspective*, New York: Guilford Press.

GRAHAM, I., 1994, *Object-Oriented Methods*, London: Addison-Wesley.

HÄGERSTRAND, T., 1975, Space, time and human conditions, in KARLQVIST, A., LUNQVIST, L. and SNICKARS, F. (eds) *Dynamic Allocation of Urban Space*, Farnborough: Saxon House, pp. 3–14.

HALL, E., 1966, *The Hidden Dimension*, London: Bodley Head.

HARVEY, D., 1969, *Explanation in Geography*, New York: St Martin's Press.

HAZELTON, N.W.J., LEAHY, F.J. and WILLIAMSON, I.P., 1990, On the design of temporally referenced 3-D geographic information systems: Development of four-dimensional GIS, *Proceedings of the GIS/LIS'90 Conference*, pp. 357–72.

JACKSON, R.W., 1994, Object-oriented modeling in regional science: an advocacy view. *Papers in Regional Science*, **73**(4), 1–21.

JACOBSON, I., CHRISTERSON, M., JONSSON, P. and OVERGARRD, G., 1992, *Object-Oriented Software Engineering*, Workingham: Addison Wesley.

JAMMER, M., 1969, *Concepts of Space*, Cambridge MA: Harvard University Press.

JOHNSON, D. and KEMP, Z., 1995, Enhancing a GIS with temporal capabilities, *Proceedings of GISRUK'95*, Extended Abstracts, pp. 25–6.

JONES, S. and MASON, P.J., 1980, Handling the time dimension in a data base, *Proceedings of the International Conference on Data Bases*, pp. 65–83.

JONES, S., MASON, P.J. and STAMPER, R., 1979, LEGOL 2.0: a relational specification language for complex rules. *Information Systems*, **4**(4), 293–305.

JONES, S.B., 1945, *Boundary-Making: A Handbook for Statesmen, Treaty Editors and Boundary Commissioners*, Washington DC: Carnegie Endowment for International Peace.

KARLQVIST, A., LUNQVIST, L. and SNICKARS, F. (eds) 1975, *Dynamic Allocation of Urban Space*, Farnborough: Saxon House.

KEMP, Z. and KOWALCZYK, A., 1994, Incorporating the temporal dimension in a geographical information system, in WORBOYS, M.F. (ed.) *Innovations in GIS*, London: Taylor & Francis.

KIM, W., 1991, Object-oriented database systems: strengths and weakness. *Journal of Object-Oriented Programming*, **July/Aug**, 21–3.

KIM, W. and LOCHOVSKY, F.H., 1989, *Object-Oriented Concepts, Applications and Databases*, Reading MA: Addison-Wesley.

KRAAK, M. and MACEACHREN, A.M., 1994, Visualization of the temporal component of spatial data, *Proceedings of the SDH'94 Conference*, Vol. 1, pp. 391–409.

KRASNER, G., 1981, The Smalltalk-80 virtual machine. *Byte*, **6**(8), 12–20.

KUCERA, G.L., 1996, *Temporal Extensions to Spatial Data Models: Final Report*, US Army Construction Engineering Research Laboratory.

LANGRAN, G., 1988, Temporal GIS design tradeoffs, *Proceedings of the GIS/LIS'88 Conference*, pp. 890–99.

LANGRAN, G., 1989, A review of temporal database research and its use in GIS applications. *International Journal of Geographical Information Systems*, **3**(3), 215–32.

LANGRAN, G., 1992a, *Time in Geographic Information Systems*, London: Taylor & Francis.

LANGRAN, G., 1992b, States, events, and evidence: the principle entities of a temporal GIS, *Proceedings of the GIS/LIS'92 Conference*, Vol. 1, pp. 416–25.

LANGRAN, G., 1993, Issues of implementing a spatiotemporal system. *International Journal of Geographical Information Systems*, **7**(4), 305–14.

LAPRADELLE, P. DE, 1928, *La Frontière: Etude de Droit International* (The Boundary: A Study of International Law), Paris: Les Editions Internationales.

LAURINI, R. and THOMPSON, D., 1992, *Fundamentals of Spatial Information Systems*, San Diego CA: Academic Press.

LENNTORP, B., 1976, *Paths in Space-Time Environments: A Time-Geographic Study of Movement Possibilities of Individuals*, Lund: Royal University of Lund.

LENNTORP, B., 1978, A time-geographic simulation model of individual activity programmes, in CARLSTEIN, T., PARKES, D. and THRIFT, N. (eds) *Timing Space and Spacing Time*, Vol. 2, *Human Activity and Time Geography*, Lund: Royal University of Lund, pp. 162–80.

LOOMIS, M.E.S., 1992, Object versioning. *Journal of Object-Oriented Programming*, **Jan**, 40–43.

LUM, V.P., DADAM, P., ERBE, R., GUENAUER, J., PISTOR, P., WALCH, G., WERNER, H. and WOODFILL, J., 1984, Designing DBMS support for the temporal dimension, *Proceedings of the ACM SIGMOD Conference on Management of Data*, pp. 115–30.

MCCORMICK, B.H., DEFANTI, T.A. and BROWN, M.D., 1987, Visualisation in scientific computing. *SIGGRAPH Computer Graphics Newsletter*, **21**(6).

MACEACHREN, A.M., 1995, *How Maps Work: Representation, Visualization, and Design*, New York: Guilford Press.

MACEACHREN, A.M. and TAYLOR, D.R.F., 1998, *Visualization in Modern Cartography*, London: Pergamon.

MACEACHREN, A.M., WACHOWICZ, M., EDSALL, R., HAUG, D. and MASTERS, R., 1999, Constructing knowledge from multivariate spatiotemporal data: integrating GVis with KDD methods. *International Journal of Geographical Information Sciences*, in Press.

MAKIN, J., 1992, An object-oriented simulation of a complex geographical system using GIS, MSc dissertation (unpublished), University of Edinburgh.

MÅRTENSSON, S., 1978, Time allocation and daily living conditions: comparing regions, in CARLSTEIN, T., PARKES, D. and THRIFT, N. (eds) *Timing Space and Spacing Time*, Vol. 2, *Human Activity and Time Geography*, Lund: Royal University of Lund, pp. 181–97.

MILLER, H.J., 1991, Modelling accessibility using space-time prism concepts within geographical information systems. *International Journal of Geographical Information Systems*, **5**(3), 287–301.

MILNE, P., MILTON, S. and SMITH, J., 1993, Geographical object-oriented databases – a case study. *International Journal of Geographical Information Systems*, **7**(1), 39–55.

MUELLER, T. and STEINBAUER, D., 1983, Eine Sprachschnittstele zur Versionenkontrolle in CAM-Datenbaken, in *Informatik-Fachberichte*, Berlin: Springer-Verlag, pp. 76–95.

NEWELL, R.G. and BATTY, P.M., 1993, GIS databases are different, *Proceedings of the AGI'93 Conference*, pp. 3.2.1–3.2.4.

NEWELL, R.G., THERIAULT, D.G. and EASTERFIELD, M., 1994, Temporal GIS – modelling the evolution of spatial data in time. *Smallworld Technical Report*, Paper 6.

NIST, 1991, *X3/SPARC/DBSSG/OODBTG: Final Technical Report of the American National Standards Institute*. Gaithersburg MD: National Institute of Standards and Technology.

ODMG, 1994, Response to the March 1994 ODMG-93 commentary. *SIGMOD Record*, **23**(3), 3–7.

OLANDER, L. and CARLSTEIN, T., 1978, The study of activities in the quaternary sector, in CARLSTEIN, T., PARKES, D. and THRIFT, N. (eds) *Timing Space and Spacing Time*, Vol. 2, *Human Activity and Time Geography*, Lund: Royal University of Lund, pp. 198–213.

ORNSTEIN, R.E., 1969, *On the Experience of Time*, London: Penguin.

PARKES, D. and THRIFT, N., 1980, *Times, Spaces, and Places: A Chronogeography Perspective*, Chichester: John Wiley.

PEUQUET, D., 1994, It's about time: a conceptual framework for the representation of temporal dynamics in geographic information systems. *Annals of the Association of American Geographers*, **84**(3), 441–61.

PEUQUET, D. and WENTZ, E., 1994, An approach for time-based analysis of spatiotemporal data. *Proceedings of the SDH'94 Conference*, Vol. 1, pp. 489–504.

PRED, A., 1977, The choreography of existence: comments on Hägerstrand's time geography and its usefulness. *Economic Geography*, **53**, 207–221.

PRESCOTT, J.R.V., 1987, *Political Frontiers and Boundaries*, London: Unwin Hyman.

QIAN, L., WACHOWICZ, M., PEUQUET, D. and MACEACHREN, A.M., 1997, Data processing operations for visualization and analysis of space-time data in GIS. *Proceedings of GIS/LIS'97.*

RACKHAM, L.J., 1987, The creation of a prototype relational database for public boundaries and administrative areas in Scotland, MSc dissertation (unpublished), University of Edinburgh.

RACKHAM, L.J., 1992, Development of a system for the management and supply of data on administrative areas and public boundaries, *Updating of Digital Maps and Topographic Databases*, Proceedings of the third meeting of CERCO Working Group IX, pp. 1–13.

RAMACHANDRAN, B., 1992, Modelling temporal changes in the structure of real-world entities within a GIS environment using an object-oriented approach, MSc dissertation (unpublished), University of Edinburgh.

REED, D., 1978, Naming and synchronization in a decentralized computer system, PhD dissertation (unpublished), MIT.

RENOLEN, A., 1996, History graphs: conceptual modelling of spatiotemporal data, *Proceedings of Brno GIS Conference.*

ROJAS-VEGA, E. and KEMP, Z., 1994, Object-orientation and spatial data modelling: a formal approach. Poster Session at the UKRGIS'94 Conference.

RUBENSTEIN, R. and HERSH, H., 1984, *The Human Factor: Designing Computer Systems for People*, Bedford TX: Digital Press.

RUMBAUGH, J., BLAHA, M., PREMERLANI, W., EDDY, F. and LORENSEN, W., 1991, *Object-Oriented Modelling and Design*, Englewood Cliffs NJ: Prentice Hall.

SCHNEIDER, R. and KRIEGEL, H.P., 1992, Indexing the spatio-temporal monitoring of a polygon object, *Proceedings of the SDH'92 Conference*, Vol. 1, pp. 209–20.

SHLAER, S. and MELLOR, S.J., 1988, *Objected-Oriented Systems Analysis: Modeling the World in Data*, Englewood Cliffs NJ: Prentice Hall.

SHOHAM, Y. and GOYAL, N., 1988, Temporal reasoning in artificial intelligence, in SHROBE, H.E. and the American Association for Artificial Intelligence (eds) *Exploring Artificial Intelligence: Survey Talks from the National Conferences on Artificial Intelligence*, San Mateo CA: Morgan Kaufmann.

SNODGRASS, R.T., 1987, The temporal query language TQuel. *ACM Transactions on Database Systems*, **12**(2), 247–98.

SNODGRASS, R.T., 1990, Temporal databases: status and research directions. *SIGMOD Record*, **19**(4), 83–9.

SNODGRASS, R.T., 1992, Temporal databases, in FRANK, A.U., CAMPARI, I. and FORMENTINI, U. (eds) *Theories and Methods of Spatio-Temporal Reasoning in Geographic Space*, London: Springer-Verlag, pp. 22–64.

SNODGRASS, R.T. and AHN, I., 1985, A taxonomy of time in databases, *Proceedings of the ACM-SIGMOD Conference on Management of Data*, pp. 236–46.

SNODGRASS, R.T. and AHN, I., 1986, Temporal databases. *Computer*, **19**(9), 35–42.

SNODGRASS, R.T., AL-TAHA, K. and SOO, M.D., 1993, Bibliography on spatiotemporal databases. *SIGMOD Record*, **17**(1), 10–21.

SOO, M.D., 1991, Bibliography on temporal databases. *SIGMOD Record*, **20**(1), 14–23.

STONEBRAKER, M., 1987, The design of the POSTGRES storage system, *Proceedings of the Very Large Databases Conference*, pp. 289–300.

STONEBRAKER, M. and MOORE, M., 1996, *Object-Relational DBMS: The Next Great Wave*, San Francisco CA: Morgan Kaufmann.

STROUSTRUP, B., 1988, What is object-oriented programming? *IEEE Software*, **May**, 10–20.

SVENSSON, P. and HUANG, Z., 1991, Geo-SAL: a query language for spatial data analysis, *Proceedings of SSD'91*, pp. 119–40.

TANSEL, A.U., CLIFFORD, J., GADIA, S., JAJODIA, S., SEGEV, A. and SNODGRASS, R., 1993, *Temporal Databases – Theory, Design, and Implementation.* Redwood City CA: Benjamin/Cummings.

THEWESSEN, T., VAN DE VELDE, R. and VERLOUW H., 1992, European groundwater threats analyzed with GIS. *GIS Europe*, **1**(3), 28–33.

VAN HOOP, S. and VAN OOSTEROM, P., 1992, Storage and manipulation of topology in POSTGRES, *Proceedings of the EGIS'92 Conference*, Vol. 2, pp. 1324–36.

VERBURG, P.H., KONING, G.H.J., KOK, K., VELDKAMP, A., FRESCO, L.O. and BOUMA, J., 1997, Quantifying the spatial structure of land use change: an integrated approach data, *Proceedings of the International Conference on Geo-Information for Sustainable Land Management*, pp. 1–9.

WACHOWICZ, M. and BROADGATE, M.L., 1993, A significant challenge: prediction of environmental changes using a temporal GIS, *Proceedings of the AGI'93 Conference*, pp. 2.25.1–2.25.5.

WACHOWICZ, M., PEUQUET, D.J. and MACEACHREN, A.M., 1998, Integrating data mining and GVis for exploring spatio-temporal data, *Proceedings of GISRUK'98*.

WASSERMAN, A.I., PIRCHER, P.A. and MULLER, R.J., 1990, The object-oriented structure design for software design representation. *IEEE Computer*, **Mar**, 50–62.

WEGNER, P. and ZDONIK, S.B., 1988, Inheritance as an incremental modification mechanism or what like is and isn't like, *Proceedings of ECOOP'88*, pp. 55–7.

WORBOYS, M.F., 1994, Unifying the spatial and temporal components of geographical information, *Proceedings of the SDH'94 Conference*, Vol. 1, pp. 505–517.

WORBOYS, M.F., HEARNSHAW, H. and MAGUIRE, D., 1990, Object-oriented data modelling for spatial databases. *International Journal of Geographical Information Systems*, **4**(4), 369–83.

Index

![I]

Printed and bound by CPI Group (UK) Ltd, Croydon, CR0 4YY

24/10/2024

01778302-0002